CHICAGO PUBLIC LIBRARY
P9-CLC-065

Chicago Public Library
C
P
L
REFERENCE
Form 178 rev. 11-00

DISCARD

career ideas for teens in manufacturing

Diane Lindsey Reeves
with Gail Karlitz

Ferguson
An imprint of Facts On File

Career Ideas for Teens in Manufacturing

Ferguson
An imprint of Facts On File, Inc.
132 West 31st Street
New York NY 10001

Library of Congress Cataloging-in-Publication Data

Reeves, Diane Lindsey, 1959–
Career ideas for teens in manufacturing / Diane Lindsey Reeves with Gail Karlitz.
p. cm.
Includes index.
ISBN 0-8160-5294-8 (hc : alk. paper)
1. Manufacturing industries—Vocational guidance—Juvenile literature. I. Karlitz, Gail. II. Title.
HD9720.5.R44 2004
670'.23—dc22

2004010501

Ferguson books are available at special discounts when purchased in bulk quantities for businesses, associations, institutions, or sales promotions. Please call our Special Sales Department in New York at (212) 967-8800 or (800) 322-8755.

You can find Ferguson on the World Wide Web at http://www.fergpubco.com

Text design by Joel and Sandy Armstrong
Cover design by Nora Wertz
Cover illustration by Matt Wood
Illustrations by Jim Golden and Matt Wood

Printed in the United States of America

VB PKG 10 9 8 7 6 5 4 3 2 1

This book is printed on acid-free paper.

contents

acknowledgments

A million thanks to the people who took the time to share their career stories and provide photos for this book:

Mark Fitzgerald
Diana Gubitosi
Aly Khalifa
Nichol Mackey
John Potter, Jr.
Ray Rocker
Dale Senatore

and to Anna Prokos for content contributions.

career ideas for teens

welcome to your future

Q: What's one of the most boring questions adults ask teens?

A: "So . . . what do you want to be when you grow up?"

Well-meaning adults always seem so interested in what you plan to be.

You, on the other hand, are just trying to make it through high school in one piece.

But you may still have a nagging feeling that you really need to find some direction and think about what you want to do with your life.

When it comes to choosing your life's work there's some good news and some bad news. The good news is that, according to the U.S. Bureau of Labor Statistics, you have more than 12,000 different occupations to choose from. With that many options there's got to be something that's just right for you.

Right?

Absolutely.

But . . .

Here comes the bad news.

THERE ARE MORE THAN 12,000 DIFFERENT OCCUPATIONS TO CHOOSE FROM!

How in the world are you ever going to figure out which one is right for you?

We're so glad you asked!

Helping high school students like you make informed choices about their future is what this book (and each of the other titles in the *Career Ideas for Teens* series) is all about. Here you'll encounter 10 tough questions designed to help you answer the biggest one of all: "What in the world am I going to do after I graduate from high school?"

The *Career Ideas for Teens* series enables you to expand your horizons beyond the "doctor, teacher, lawyer" responses common to those new to the career exploration process. The books provide a no-pressure introduction to real jobs that real people do. And they offer a chance to "try on" different career options before committing to a specific college program or career path. Each title in this series is based on one of the 16 career clusters established by the U.S. Department of Education.

And what is a career cluster, you ask? Career clusters are based on a simple and very useful concept. Each cluster consists of all entry-level through professional-level occupations in a broad industry area. All of the jobs and industries in a cluster have many things in common. This organizational structure makes it easier for people like you to get a handle on the big world of work. So instead of rushing headlong into a mind-boggling exploration of the entire universe of career opportunities, you get a chance to tiptoe into smaller, more manageable segments first.

We've used this career cluster concept to organize the *Career Ideas for Teens* series of books. For example, careers related to the arts, communication, and entertainment are organized or "clustered" into the *Career Ideas for Teens in the Arts and Communications* title; a wide variety of health care professions are included in *Career Ideas for Teens in Health Science*; and so on.

Clueless as to what some of these industries are all about? Can't even imagine how something like manufacturing or public administration could possibly relate to you?

No problem.

You're about to find out. Just be prepared to expect the unexpected as you venture out into the world of work. There are some pretty incredible options out there, and some pretty surprising ones too. In fact, it's quite possible that you'll discover that the ideal career for you is one you had never heard of before.

Whatever you do, don't cut yourself short by limiting yourself to just one book in the series. You may find that your initial interests guide you towards the health sciences field—which would, of course, be a good place to start. However, you may discover some new "twists" with a look through the arts and communications book. There you may find a way to blend your medical interests with your exceptional writing and speaking skills by considering becoming a public relations (PR) specialist for a hospital or pharmaceutical company. Or look at the book on education to see about becoming a public health educator or school nurse.

Before you get started, you should know that this book is divided into three sections, each representing an important step toward figuring out what to do with your life.

The first eight titles in the *Career Ideas for Teens* series focus on:

- Architecture and Construction
- Arts and Communications
- Education and Training
- Government and Public Service
- Health Science
- Information Technology
- Law and Public Safety
- Manufacturing

Before You Get Started

Unlike most books, this one is meant to be actively experienced, rather than merely read. Passive perusal won't cut it. Energetic engagement is what it takes to figure out something as important as the rest of your life.

As we've already mentioned, you'll encounter 10 important questions as you work your way through this book. Following each Big Question is an activity designated with a symbol that looks like this:

Every time you see this symbol, you'll know it's time to invest a little energy in your future by getting out your notebook or binder, a pen or pencil, and doing whatever the instructions direct you to do. If this book is your personal property, you can choose to do the activities right in the book. But you still might want to make copies of your finished products to go in a binder so they are all in one place for easy reference.

When you've completed all the activities, you'll have your own personal **Big Question AnswerBook**, a planning guide representing a straightforward and truly effective process you can use throughout your life to make fully informed career decisions.

discover you at work

This first section focuses on a very important subject: You. It poses four Big Questions that are designed specifically to help you "discover you":

- Big Question #1: **who are you?**
- Big Question #2: **what are your interests and strengths?**
- Big Question #3: **what are your work values?**

Then, using an interest assessment tool developed by the U.S. Department of Labor and implemented with your very vivid imagination, you'll picture yourself doing some of the things that people actually do for their jobs. In other words, you'll start "discovering you at work" by answering the following:

- Big Question #4: **what's your work personality?**

Unfortunately, this first step is often a misstep for many people. Or make that a "missed" step. When you talk with the adults in your life about their career choices, you're likely to find that some of them never even considered the idea of choosing a career based on personal preferences and strengths. You're also likely to learn that if they had it to do over again, this step would definitely play a significant role in the choices they would make.

explore your options

There's more than meets the eye when it comes to finding the best career to pursue. There are also countless ways to blend talent or passion in these areas in some rather unexpected and exciting ways. Get ready to find answers to two more Big Questions as you browse through an entire section of career profiles:

? Big Question #5: **do you have the right skills?**
? Big Question #6: **are you on the right path?**

experiment with success

At long last you're ready to give this thing called career planning a trial run. Here's where you'll encounter three Big Questions that will unleash critical decision-making strategies and skills that will serve you well throughout a lifetime of career success.

While you're at it, take some time to sit in on a roundtable discussion with successful professionals representing a very impressive array of careers related to this industry. Many of their experiences will apply to your own life, even if you don't plan to pursue the same careers.

? Big Question #7: **who knows what you need to know?**
? Big Question #8: **how can you find out what a career is really like?**
? Big Question #9: **how do you know when you've made the right choice?**

Then as you begin to pull all your new insights and ideas together, you'll come to one final question:

? Big Question #10: **what's next?**

As you get ready to take the plunge, remember that this is a book about possibilities and potential. You can use it to make the most of your future work!

Here's what you'll need to complete the Big Question AnswerBook:

- A notebook or binder for the completed activities included in all three sections of the book
- An openness to new ideas
- Complete and completely candid answers to the 10 Big Question activities

So don't just read it, do it.
Plan it.
Dream it.

SECTION 1 discover you at work

The goal here is to get some clues about who you are and what you should do with your life. As time goes by, you will grow older, become more educated, and have more experiences, but many things that truly define you are not likely to change. Even now you possess very strong characteristics—genuine qualities that mark you as the unique and gifted person that you undoubtedly are.

It's impossible to overestimate the importance of giving your wholehearted attention to this step. You, after all, are the most valuable commodity you'll ever have to offer a future employer. Finding work that makes the most of your assets often means the difference between enjoying a rewarding career and simply earning a paycheck.

You've probably already experienced the satisfaction of a good day's work. You know what we mean—those days when you get all your assignments in on time, you're prepared for the pop quiz your teacher sprung on you, and you beat your best time during sports practice. You may be exhausted at the end of the day but you can't help but feel good about yourself and your accomplishments. A well-chosen career can provide that same sense of satisfaction. Since you're likely to spend upwards of 40 years doing some kind of work, well-informed choices make a lot of sense!

Let's take a little time for you to understand yourself and connect what you discover about yourself to the world of work.

To find a career path that's right for you, we'll tackle these three Big Questions first:

- **who are you?**
- **what are your interests and strengths?**
- **what are your work values?**

? Big Question #1: who are you?

Have you ever noticed how quickly new students in your school or new families in your community find the people who are most like them? If you've ever been the "new" person yourself, you've probably spent your first few days sizing up the general population and then getting in with the people who dress in clothes a lot like yours, appreciate the same style of music, or maybe even root for the same sports teams.

Given that this process happens so naturally—if not necessarily on purpose—it should come as no surprise that many people lean toward jobs that surround them with people most like them. When people with common interests, common values, and complementary talents come together in the workplace, the results can be quite remarkable.

Many career aptitude tests, including the one developed by the U.S. Department of Labor and included later in this book, are based on the theory that certain types of people do better at certain types of jobs. It's like a really sophisticated matchmaking service. Take your basic strengths and interests and match them to the strengths and interests required by specific occupations.

It makes sense when you think about it. When you want to find a career that's ideally suited for you, find out what people like you are doing and head off in that direction!

There's just one little catch.

The only way to recognize other people like you is to recognize yourself. Who are you anyway? What are you like? What's your basic approach to life and work?

Now's as good a time as any to find out. Let's start by looking at who you are in a systematic way. This process will ultimately help you understand how to identify personally appropriate career options.

Big Activity #1:
who are you?

On a sheet of paper, if this book doesn't belong to you, create a list of adjectives that best describe you. You should be able to come up with at least 15 qualities that apply to you. There's no need to make judgments about whether these qualities are good or bad. They just are. They represent who you are and can help you understand what you bring to the workforce.

(If you get stuck, ask a trusted friend or adult to help describe especially strong traits they see in you.)

Some of the types of qualities you may choose to include are:

- **How you relate to others:**
  Are you shy? Outgoing? Helpful? Dependent? Empathic? In charge? Agreeable? Challenging? Persuasive? Popular? Impatient? A loner?
- **How you approach new situations:**
  Are you adventurous? Traditional? Cautious? Enthusiastic? Curious?
- **How you feel about change—planned or unplanned:**
  Are you resistant? Adaptable? Flexible? Predictable?
- **How you approach problems:**
  Are you persistent? Spontaneous? Methodical? Creative?
- **How you make decisions:**
  Are you intuitive? Logical? Emotional? Practical? Systematic? Analytical?
- **How you approach life:**
  Are you laid back? Ambitious? Perfectionist? Idealistic? Optimistic? Pessimistic? Self-sufficient?

Feel free to use any of these words if they happen to describe you well, but please don't limit yourself to this list. Pick the best adjectives that paint an accurate picture of the real you. Get more ideas from a dictionary or thesaurus if you'd like.

When you're finished, put the completed list in your Big Question AnswerBook.

Big Activity #1: **who are you?**

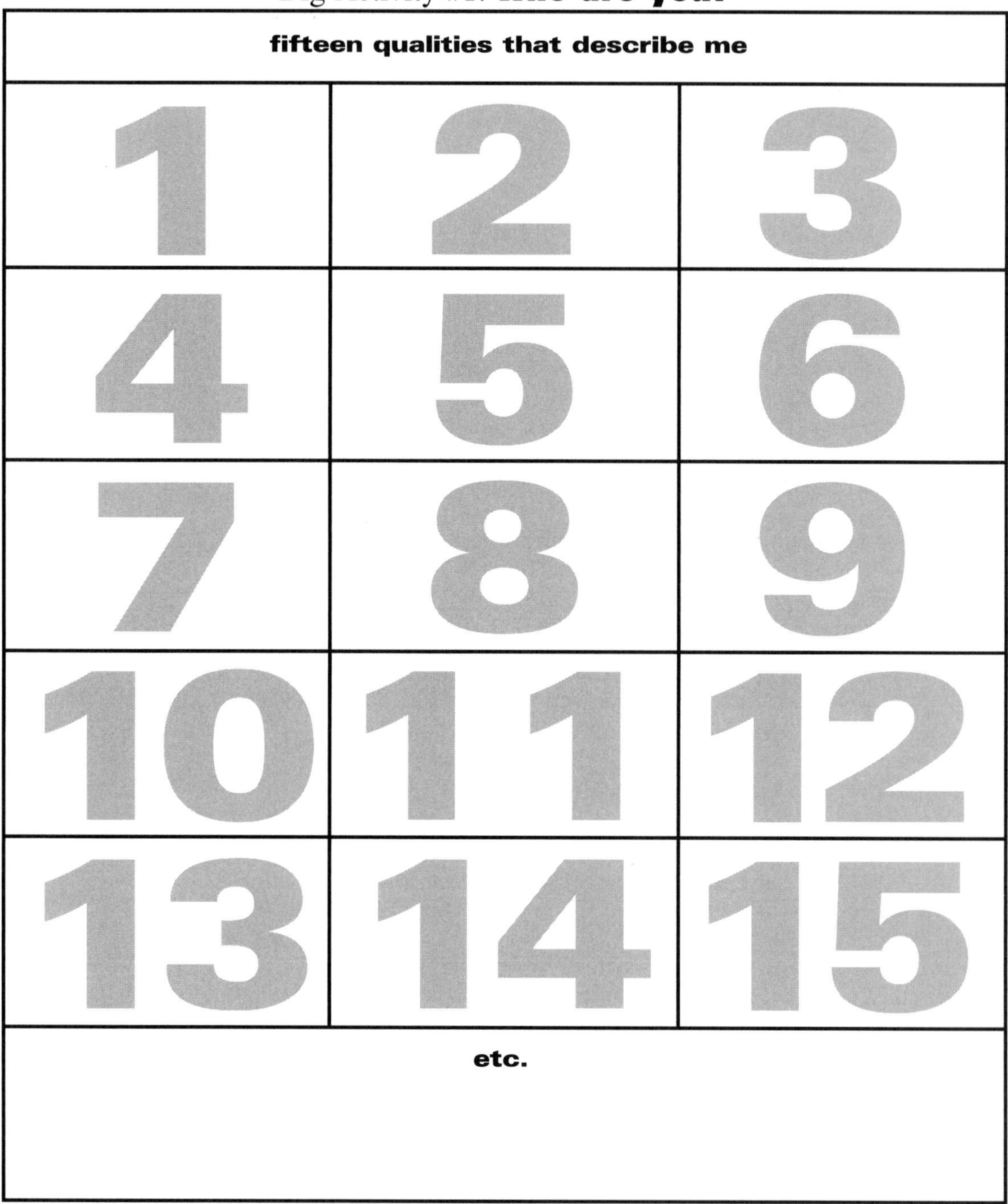

fifteen qualities that describe me		
1	2	3
4	5	6
7	8	9
10	11	12
13	14	15
etc.		

Big Question #2: what are your interests and strengths?

For many people, doing something they like to do is the most important part of deciding on a career path—even more important than how much money they can earn!

We don't all like to do the same things—and that's good. For some people, the ideal vacation is lying on a beach, doing absolutely nothing; others would love to spend weeks visiting museums and historic places. Some people wish they had time to learn to skydive or fly a plane; others like to learn to cook gourmet meals or do advanced math.

If we all liked the same things, the world just wouldn't work very well. There would be incredible crowds in some places and ghost towns in others. Some of our natural resources would be overburdened; others would never be used. We would all want to eat at the same restaurant, wear the same outfit, see the same movie, and live in the same place. How boring!

So let's get down to figuring out what you most like to do and how you can spend your working life doing just that. In some ways your answer to this question is all you really need to know about choosing a career, because the people who enjoy their work the most are those who do something they enjoy. We're not talking rocket science here. Just plain old common sense.

Big Activity # 2:
what are your interests and strengths?

Imagine this: No school, no job, no homework, no chores, no obligations at all. All the time in the world you want to do all the things you like most. You know what we're talking about—those things that completely grab your interest and keep you engrossed for hours without your getting bored. Those kinds of things you do really well—sometimes effortlessly, sometimes with extraordinary (and practiced) skill.

And, by the way, EVERYONE has plenty of both interests and strengths. Some are just more visible than others.

Step 1: Write the three things you most enjoy doing on a sheet of paper, if this book doesn't belong to you. Leave lots of space after each thing.

Step 2: Think about some of the deeper reasons why you enjoy each of these activities—the motivations beyond "it's fun." Do you enjoy shopping because it gives you a chance to be with your friends? Because it allows you to find new ways to express your individuality? Because you enjoy the challenge of finding bargains or things no one else has discovered? Or because it's fun to imagine the lifestyle you'll be able to lead when you're finally rich and famous? In the blank spaces, record the reasons why you enjoy each activity.

Step 3: Keep this list handy in your Big Question AnswerBook so that you can refer to it any time you have to make a vocational decision. Sure, you may have to update the list from time to time as your interests change. But one thing is certain. The kind of work you'll most enjoy will be linked in some way to the activities on that list. Count on it.

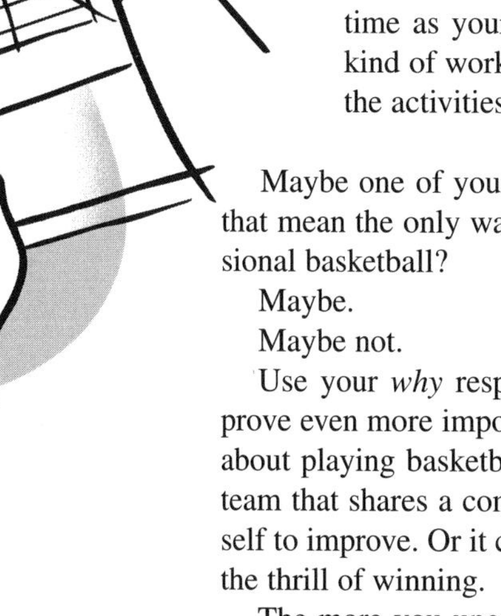

Maybe one of your favorite things to do is "play basketball." Does that mean the only way you'll ever be happy at work is to play professional basketball?

Maybe.

Maybe not.

Use your *why* responses to read between the lines. The *why*s can prove even more important than the *what*s. Perhaps what you like most about playing basketball is the challenge or the chance to be part of a team that shares a common goal. Maybe you really like pushing yourself to improve. Or it could be the rush associated with competition and the thrill of winning.

The more you uncover your own *why*s, the closer you'll be to discovering important clues about the kinds of work that are best for you.

Big Activity #2: **what are your interests and strengths?**

things you enjoy doing	why you enjoy doing them
1	• • •
2	• • •
3	• • •

? Big Question #3: what are your work values?

Chances are, you've never given a moment's thought to this next question. At least not in the context of career planning.

You already looked at who you are and what you enjoy and do well. The idea being, of course, to seek out career options that make the most of your innate qualities, preferences, and natural abilities.

As you start checking into various careers, you'll discover one more dimension associated with making personally appropriate career choices. You'll find that even though people may have the exact same job title, they may execute their jobs in dramatically different ways. For instance, everyone knows about teachers. They teach things to other people. Period.

But wait. If you line up 10 aspiring teachers in one room, you may be surprised to discover how vastly different their interpretations of the job may be. There are the obvious differences, of course. One may want to teach young children; one may want to teach adults. One will focus on teaching math, while another one focuses on teaching Spanish.

Look a little closer and you'll find even greater disparity in the choices they make. One may opt for the prestige (and paycheck) of working in an Ivy League college, while another is completely committed to teaching disadvantaged children in a remote area of the Appalachian Mountains. One may approach teaching simply as a way to make a living, while another devotes almost every waking hour to working with his or her students.

These subtle but significant differences reflect what's truly important to each person. In a word, they reflect the person's values—those things that are most important to them.

People's values depend on many factors—their upbringing, their life experiences, their goals and ambitions, their religious beliefs, and, quite frankly, the way they view the world and their role in it. Very few people share exactly the same values. However, that doesn't necessarily mean that some people are right and others are wrong. It just means they have different perspectives.

Here's a story that shows how different values can be reflected in career choices.

Imagine: It's five years after college graduation and a group of college friends are back together for the first time. They catch up about their lives, their families, and their careers. Listen in on one of their reunion conversations and see if you can guess what each is doing now.

Alice: I have the best career. Every day I get the chance to help kids with special needs get a good education.

Bob: I love my career, too. It's great to know that I am making my town a safer place for everyone.

Cathy: It was tough for me to commit to more school after college. But I'm glad I did. After all I went through when my parents divorced, I'm glad I can be there to make things easier for other families.

David: I know how you feel. I'm glad I get to do something that helps companies function smoothly and keep our economy strong. Of course, you remember that I had a hard time deciding whether to pursue this career or teaching! This way I get the best of both worlds.

Elizabeth: It's great that we both ended up in the corporate world. You know that I was always intrigued by the stock market.

So exactly what is each of the five former freshman friends doing today? Have you made your guesses?

Alice is a lawyer. She specializes in education law. She makes sure that school districts provide special needs children with all of the resources they are entitled to under the law.

Bob is a lawyer. He is a prosecuting attorney and makes his town safer by ensuring that justice is served when someone commits a crime.

Cathy is a lawyer. She practices family law. She helps families negotiate separation and divorce agreements and makes sure that adoption and custody proceedings protect everyone involved. Sometimes she even provides legal intervention to protect adults or children who are in abusive situations.

David is a lawyer. He practices employment law. He helps companies set up policies that follow fair employment practices. He also gives seminars to managers, teaching them what the law says and means about sexual harassment, discrimination, and termination of employment.

Elizabeth is a lawyer. She practices corporate law and is indispensable to corporations with legal responsibilities towards stockholders and the government.

Wow! All five friends have the same job title. But each describes his/her job so differently! All five were able to enter the field of law and focus on the things that are most important to them: quality education, freedom from crime, protection of families and children, fairness in the workplace, and corporate economic growth. Identifying and honoring your personal values is an important part of choosing your life's work.

Big Activity #3:
what are your work values?

Step 1: Look at the following chart. If this book doesn't belong to you, divide a sheet of paper into the following three columns:

- **Essential**

Statements that fall into this column are very important to you. If the job doesn't satisfy these needs, you're not interested.

- **Okay**

Great if the job satisfies these needs, but you can also live without them.

- **No Way**

Statements that fall into this column represent needs that are not at all important to you or things you'd rather do without or simply couldn't tolerate.

Step 2: Look over the following list of statements representing different work values. Rewrite each statement in the appropriate column. Does the first statement represent something that is critical to you to have in your work? If so, write it in the first column. No big deal either way? Write it in the second column. Couldn't stand it? Write it in the third column. Repeat the same process for each of the value statements.

Step 3: When you're finished, place your complete work values chart in your Big Question AnswerBook.

Got it? Then get with it.

Really think about these issues. Lay it on the line. What values are so deeply ingrained in you that you'd be absolutely miserable if you had to sacrifice them for a job? Religious beliefs and political leanings fall into this category for some people.

Which ones provide room for some give and take? Things like vacation and benefits, working hours, and other issues along those lines may be completely negotiable for some people, but absolutely not for others.

Just remember, wherever you go and whatever you do, be sure that the choices you make are true to you.

Big Activity #3: **what are your work values?**

work values	essential	okay	no way
1. I can count on plenty of opportunity for advancement and taking on more responsibility.			
2. I can work to my fullest potential using all of my abilities.			
3. I would be able to give directions and instructions to others.			
4. I would always know exactly what my manager expects of me.			
5. I could structure my own day.			
6. I would be very busy all day.			
7. I would work in attractive and pleasant surroundings.			
8. My coworkers would be people I might choose as friends.			
9. I would get frequent feedback about my performance.			
10. I could continue my education to progress to an even higher level job.			
11. Most of the time I would be able to work alone.			
12. I would know precisely what I need to do to succeed at the job.			
13. I could make decisions on my own.			

Big Activity #3: **what are your work values?**

work values	essential	okay	no way
14. I would have more than the usual amount of vacation time.			
15. I would be working doing something I really believe in.			
16. I would feel like part of a team.			
17. I would find good job security and stable employment opportunities in the industry.			
18. I could depend on my manager for the training I need.			
19. I would earn lots of money.			
20. I would feel a sense of accomplishment in my work.			
21. I would be helping other people.			
22. I could try out my own ideas.			
23. I would not need to have further training or education to do this job.			
24. I would get to travel a lot.			
25. I could work the kind of hours I need to balance work, family, and personal responsibilities.			

To summarize in my own words, the work values most important to me include:

? Big Question #4: what is your work personality?

Congratulations. After completing the first three activities, you've already discovered a set of skills you can use throughout your life. Basing key career decisions on factors associated with who you are, what you enjoy and do well, and what's most important about work will help you today as you're just beginning to explore the possibilities, as well as into the future as you look for ways to cultivate your career.

Now that you've got that mastered, let's move on to another important skill. This one blends some of what you just learned about yourself with what you need to learn about the real world of work. It's a reality check of sorts as you align and merge your personal interests and abilities with those required in different work situations. At the end of this task you will identify your personal interest profile.

This activity is based on the work of Dr. John Holland. Dr. Holland conducted groundbreaking research that identified different characteristics in people. He found that he could classify people into six basic groups based on which characteristics tended to occur at the same time. He also found that the characteristics that defined the different groups of people were also characteristics that corresponded to success in different groups of occupations. The result of all that work was a classification system that identifies and names six distinct groups of people who share personal interests or characteristics and are likely to be successful in a group of clearly identified jobs.

Dr. Holland's work is respected by workforce professionals everywhere and is widely used by employers and employment agencies to help people get a handle on the best types of work to pursue.

The following Work Interest Profiler (WIP) is based on Dr. Holland's theories and was developed by the U.S. Department of Labor's Employment and Training Administration as part of an important project called O*Net. O*Net is a system used in all 50 states to provide career and employment services to thousands of people every year. It's a system you'll want to know about when it's time to take that first plunge into the world of work. If you'd like, you can find more information about this system at ***http://online.onetcenter.org***.

Big Activity #4:

what is your work personality?

By completing O*Net's Work Interest Profiler (WIP), you'll gain valuable insight into the types of work that are right for you.

here's how it works

The WIP lists many activities that real people do at real jobs. Your task is to read a brief statement about each of these activities and decide if it is something you think you'd enjoy doing. Piece of cake!

Don't worry about whether you have enough education or training to perform the activity. And, for now, forget about how much money you would make performing the activity.

Just boil it down to whether or not you'd like performing each work activity. If you'd like it, put a check in the *like* column that corresponds to each of the six interest areas featured in the test on the handy dandy chart you're about to create (or use the one in the book if it's yours). If you don't like it, put that check in the *dislike* column. What if you don't have a strong opinion on a particular activity? That's okay. Count that one as *unsure*.

Be completely honest with yourself. No one else is going to see your chart. If you check things you think you "should" check, you are not helping yourself find the career that will make you happy.

Before you start, create a chart for yourself. Your scoring sheet will have six horizontal rows and three vertical columns. Label the six rows as Sections 1 through 6, and label the three columns *like*, *dislike*, and *unsure*.

how to complete the Work Interest Profiler

Step 1: Start with Section 1.

Step 2: Look at the first activity and decide whether you would like to do it as part of your job.

Step 3: Put a mark in the appropriate column (*Like*, *Dislike*, or *Unsure*) on the Section 1 row.

Step 4: Continue for every activity in Section 1. Then do Sections 2 through 6.

Step 5: When you've finished all of the sections, count the number of marks you have in each column and write down the total.

Remember, this is not a test! There are no right or wrong answers. You are completing this profile to learn more about yourself and your work-related interests.

Also, once you've completed this activity, be sure to put your chart and any notes in your Big Question AnswerBook.

Ready? Let's go!

Section 1

1. Drive a taxi
2. Repair household appliances
3. Catch fish as a member of a fishing crew
4. Paint houses
5. Assemble products in a factory
6. Install flooring in houses
7. Perform lawn care services
8. Drive a truck to deliver packages to homes and offices
9. Work on an offshore oil-drilling rig
10. Put out forest fires
11. Fix a broken faucet
12. Refinish furniture
13. Guard money in an armored car
14. Lay brick or tile
15. Operate a dairy farm
16. Raise fish in a fish hatchery
17. Build a brick walkway
18. Enforce fish and game laws
19. Assemble electronic parts
20. Build kitchen cabinets
21. Maintain the grounds of a park
22. Operate a motorboat to carry passengers
23. Set up and operate machines to make products
24. Spray trees to prevent the spread of harmful insects
25. Monitor a machine on an assembly line

Section 2

1. Study space travel
2. Develop a new medicine
3. Study the history of past civilizations
4. Develop a way to better predict the weather
5. Determine the infection rate of a new disease
6. Study the personalities of world leaders
7. Investigate the cause of a fire
8. Develop psychological profiles of criminals
9. Study whales and other types of marine life
10. Examine blood samples using a microscope
11. Invent a replacement for sugar
12. Study genetics
13. Do research on plants or animals
14. Study weather conditions
15. Investigate crimes
16. Study ways to reduce water pollution
17. Develop a new medical treatment or procedure
18. Diagnose and treat sick animals
19. Conduct chemical experiments
20. Study rocks and minerals
21. Do laboratory tests to identify diseases
22. Study the structure of the human body
23. Plan a research study
24. Study the population growth of a city
25. Make a map of the bottom of the ocean

Section 3

1. Paint sets for a play
2. Create special effects for movies
3. Write reviews of books or movies
4. Compose or arrange music
5. Design artwork for magazines
6. Pose for a photographer
7. Create dance routines for a show
8. Play a musical instrument
9. Edit movies
10. Sing professionally
11. Announce a radio show
12. Perform stunts for a movie or television show
13. Design sets for plays
14. Act in a play
15. Write a song
16. Perform jazz or tap dance
17. Sing in a band
18. Direct a movie
19. Write scripts for movies or television shows
20. Audition singers and musicians for a musical show
21. Conduct a musical choir
22. Perform comedy routines in front of an audience
23. Dance in a Broadway show
24. Perform as an extra in movies, plays, or television shows
25. Write books or plays

Section 4

1. Teach children how to play sports
2. Help people with family-related problems
3. Teach an individual an exercise routine
4. Perform nursing duties in a hospital
5. Help people with personal or emotional problems
6. Teach work and living skills to people with disabilities
7. Assist doctors in treating patients
8. Work with juveniles on probation
9. Supervise the activities of children at a camp
10. Teach an elementary school class
11. Perform rehabilitation therapy
12. Help elderly people with their daily activities
13. Help people who have problems with drugs or alcohol
14. Teach a high school class
15. Give career guidance to people
16. Do volunteer work at a non-profit organization
17. Help families care for ill relatives
18. Teach sign language to people with hearing disabilities
19. Help people with disabilities improve their daily living skills
20. Help conduct a group therapy session
21. Work with children with mental disabilities
22. Give CPR to someone who has stopped breathing
23. Provide massage therapy to people
24. Plan exercises for patients with disabilities
25. Counsel people who have a life-threatening illness

Section 5

1. Sell CDs and tapes at a music store
2. Manage a clothing store
3. Sell houses
4. Sell computer equipment in a store
5. Operate a beauty salon or barber shop
6. Sell automobiles
7. Represent a client in a lawsuit
8. Negotiate business contracts
9. Sell a soft drink product line to stores and restaurants
10. Start your own business
11. Be responsible for the operations of a company
12. Give a presentation about a product you are selling
13. Buy and sell land
14. Sell restaurant franchises to individuals
15. Manage the operations of a hotel
16. Negotiate contracts for professional athletes
17. Sell merchandise at a department store
18. Market a new line of clothing
19. Buy and sell stocks and bonds
20. Sell merchandise over the telephone
21. Run a toy store
22. Sell hair-care products to stores and salons
23. Sell refreshments at a movie theater
24. Manage a retail store
25. Sell telephone and other communication equipment

Section 6

1. Develop an office filing system
2. Generate the monthly payroll checks for an office
3. Proofread records or forms
4. Schedule business conferences
5. Enter information into a database
6. Photocopy letters and reports
7. Keep inventory records
8. Record information from customers applying for charge accounts
9. Load computer software into a large computer network
10. Use a computer program to generate customer bills
11. Develop a spreadsheet using computer software
12. Operate a calculator
13. Direct or transfer office phone calls
14. Use a word processor to edit and format documents
15. Transfer funds between banks, using a computer
16. Compute and record statistical and other numerical data
17. Stamp, sort, and distribute office mail
18. Maintain employee records
19. Record rent payments
20. Keep shipping and receiving records
21. Keep accounts payable/receivable for an office
22. Type labels for envelopes and packages
23. Calculate the wages of employees
24. Take notes during a meeting
25. Keep financial records

Section 1
Realistic

	Like	Dislike	Unsure
1.			
2.			
3.			
4.			
5.			
6.			
7.			
8.			
9.			
10.			
11.			
12.			
13.			
14.			
15.			
16.			
17.			
18.			
19.			
20.			
21.			
22.			
23.			
24.			
25.			

Total Realistic

Section 2
Investigative

	Like	Dislike	Unsure
1.			
2.			
3.			
4.			
5.			
6.			
7.			
8.			
9.			
10.			
11.			
12.			
13.			
14.			
15.			
16.			
17.			
18.			
19.			
20.			
21.			
22.			
23.			
24.			
25.			

Total Investigative

Section 3
Artistic

	Like	Dislike	Unsure
1.			
2.			
3.			
4.			
5.			
6.			
7.			
8.			
9.			
10.			
11.			
12.			
13.			
14.			
15.			
16.			
17.			
18.			
19.			
20.			
21.			
22.			
23.			
24.			
25.			

Total Artistic

Section 4
Social

	Like	Dislike	Unsure
1.			
2.			
3.			
4.			
5.			
6.			
7.			
8.			
9.			
10.			
11.			
12.			
13.			
14.			
15.			
16.			
17.			
18.			
19.			
20.			
21.			
22.			
23.			
24.			
25.			

Total Social

Section 5 Enterprising

	Like	Dislike	Unsure
1.			
2.			
3.			
4.			
5.			
6.			
7.			
8.			
9.			
10.			
11.			
12.			
13.			
14.			
15.			
16.			
17.			
18.			
19.			
20.			
21.			
22.			
23.			
24.			
25.			

Total Enterprising

Section 6 Conventional

	Like	Dislike	Unsure
1.			
2.			
3.			
4.			
5.			
6.			
7.			
8.			
9.			
10.			
11.			
12.			
13.			
14.			
15.			
16.			
17.			
18.			
19.			
20.			
21.			
22.			
23.			
24.			
25.			

Total Conventional

What are your top three work personalities? List them here if this is your own book or on a separate piece of paper if it's not.

1. __________
2. __________
3. __________

all done? let's see what it means

Be sure you count up the number of marks in each column on your scoring sheet and write down the total for each column. You will probably notice that you have a lot of *like*s for some sections, and a lot of *dislike*s for other sections. The section that has the most *like*s is your primary interest area. The section with the next highest number of *like*s is your second interest area. The next highest is your third interest area.

Now that you know your top three interest areas, what does it mean about your work personality type? We'll get to that in a minute, but first we are going to answer a couple of other questions that might have crossed your mind:

- What is the best work personality to have?
- What does my work personality mean?

First of all, there is no "best" personality in general. There is, however, a "best" personality for each of us. It's who we really are and how we feel most comfortable. There may be several "best" work personalities for any job because different people may approach the job in different ways. But there is no "best work personality."

Asking about the "best work personality" is like asking whether the "best" vehicle is a sports car, a sedan, a station wagon, or a sports utility vehicle. It all depends on who you are and what you need.

One thing we do know is that our society needs all of the work personalities in order to function effectively. Fortunately, we usually seem to have a good mix of each type.

So, while many people may find science totally boring, there are many other people who find it fun and exciting. Those are the people who invent new technologies, who become doctors and researchers, and who turn natural resources into the things we use every day. Many people may think that spending a day with young children is unbearable, but those who love that environment are the teachers, community leaders, and museum workers that nurture children's minds and personalities.

When everything is in balance, there's a job for every person and a person for every job.

Now we'll get to your work personality. Following are descriptions of each of Dr. Holland's six work personalities that correspond to the six sections in your last exercise. You, like most people, are a unique combination of more than one. A little of this, a lot of that. That's what makes us interesting.

Identify your top three work personalities. Also, pull out your responses to the first three exercises we did. As you read about your top three work personalities, see how they are similar to the way you described yourself earlier.

Type 1

Realistic

Realistic people are often seen as the "Doers." They have mechanical or athletic ability and enjoy working outdoors.

Realistic people like work activities that include practical, hands-on problems and solutions. They enjoy dealing with plants, animals, and real-life materials like wood, tools, and machinery.

Careers that involve a lot of paperwork or working closely with others are usually not attractive to realistic people.

Who you are:

independent
reserved
practical
mechanical
athletic
persistent

What you like to do/what you do well:

build things
train animals
play a sport
fix things
garden
hunt or fish
woodworking
repair cars
refinish furniture

Career possibilities:

aerospace engineer
aircraft pilot
animal breeder
architect
baker/chef
building inspector
carpenter
chemical engineer
civil engineer
construction manager
dental assistant
detective
glazier
jeweler
machinist
oceanographer
optician
park ranger
plumber
police officer
practical nurse
private investigator
radiologist
sculptor

Type 2

Investigative

Investigative people are often seen as the "Thinkers." They much prefer searching for facts and figuring out problems mentally to doing physical activity or leading other people.

If Investigative is one of your strong interest areas, your answers to the earlier exercises probably matched some of these:

Who you are:

curious
logical
independent
analytical
observant
inquisitive

What you like to do/what you do well:

think abstractly
solve problems
use a microscope
do research
fly a plane
explore new subjects
study astronomy
do puzzles
work with a computer

Career possibilities:
aerospace engineer
archaeologist
CAD technician
chemist
chiropractor
computer programmer
coroner
dentist
electrician
ecologist
geneticist
hazardous waste technician
historian
horticulturist
management consultant
medical technologist
meteorologist
nurse practitioner
pediatrician
pharmacist
political scientist
psychologist
software engineer
surgeon
technical writer
veterinarian
zoologist

Type 3
Artistic

Artistic people are the "Creators." People with this primary interest like work activities that deal with the artistic side of things.

Artistic people need to have the opportunity for self-expression in their work. They want to be able to use their imaginations and prefer to work in less structured environments, without clear sets of rules about how things should be done.

Who you are:
imaginative
intuitive
expressive
emotional
creative
independent

What you like to do/what you do well:
draw
paint
play an instrument
visit museums
act
design clothes or rooms
read fiction
travel
write stories, poetry, or music

Career possibilities:
architect
actor
animator
art director
cartoonist
choreographer
costume designer
composer
copywriter
dancer
disc jockey
drama teacher
emcee
fashion designer
graphic designer
illustrator
interior designer
journalist
landscape architect
medical illustrator
photographer
producer
scriptwriter
set designer

Type 4 **Social**

Social people are known as the "Helpers." They are interested in work that can assist others and promote learning and personal development.

Communication with other people is very important to those in the Social group. They usually do not enjoy jobs that require a great amount of work with objects, machines, or data. Social people like to teach, give advice, help, cure, or otherwise be of service to people.

Who you are:
friendly
outgoing
empathic
persuasive
idealistic
generous

What you like to do/what you do well:
teach others
work in groups
play team sports
care for children
go to parties
help or advise others
meet new people
express yourself
join clubs or organizations

Career possibilities:
animal trainer
arbitrator
art teacher
art therapist
audiologist
child care worker
clergy person
coach
counselor/therapist
cruise director
dental hygienist
employment interviewer
EMT worker
fitness trainer
flight attendant
occupational therapist
police officer
recreational therapist
registered nurse
school psychologist
social worker
substance abuse counselor
teacher
tour guide

Type 5 **Enterprising**

Enterprising work personalities can be called the "Persuaders." These people like work activities that have to do with starting up and carrying out projects, especially business ventures. They like taking risks for profit, enjoy being responsible for making decisions, and generally prefer action to thought or analysis.

People in the Enterprising group like to work with other people. While the Social group focuses on helping other people, members of the Enterprising group are able to lead, manage, or persuade other people to accomplish the goals of the organization.

Who you are:
assertive
self-confident
ambitious
extroverted
optimistic
adventurous

What you like to do/what you do well:
organize activities
sell things
promote ideas

discuss politics
hold office in clubs
give talks or speeches
meet people
initiate projects
start your own business

Career possibilities:
advertising
chef
coach, scout
criminal investigator
economist
editor
foreign service officer
funeral director
hotel manager
journalist
lawyer
lobbyist
public relations specialist
newscaster
restaurant manager
sales manager
school principal
ship's captain
stockbroker
umpire, referee
urban planner

Type 6
Conventional

People in the Conventional group are the "Organizers." They like work activities that follow set procedures and routines. They are more comfortable and proficient working with data and detail than they are with generalized ideas.

Conventional people are happiest in work situations where the lines of authority are clear, where they know exactly what responsibilities are expected of them, and where there are precise standards for the work.

Who you are:
well-organized
accurate
practical
persistent
conscientious
ambitious

What you like to do/what you do well:
work with numbers
type accurately
collect or organize things
follow up on tasks
be punctual
be responsible for details
proofread
keep accurate records
understand regulations

Career possibilities:
accountant
actuary
air traffic controller
assessor
budget analyst
building inspector
chief financial officer
corporate treasurer
cost estimator
court reporter
economist
environmental compliance lawyer
fire inspector
insurance underwriter
legal secretary
mathematician
medical secretary
proofreader
tax preparer

manufacturing careers work personality chart

Once you've discovered your own unique work personality code, you can use it to explore the careers profiled in this book and elsewhere. Do keep in mind though that this code is just a tool meant to help focus your search. It's not meant to box you in or to keep you from pursuing any career that happens to capture your imagination.

Following is a chart listing the work personality codes associated with each of the careers profiled in this book.

	Realistic	Investigative	Artistic	Social	Enterprising	Conventional
My Work Personality Code (mark your top three areas)						
Assembler	X					
Avionics Technician	X	X				X
Boilermaker	X					
CAM (Computer-Aided Manufacturing) Technician	X		X			X
Chemical Engineer	X	X				
Computer Hardware Engineer	X	X				X
Cost Estimator	X					X
Electrical Engineer	X	X				
Electronic Equipment Assembler	X					
Environmental Engineer	X	X				X
Foundry Worker	X					X
Hoist and Winch Operator	X					X
Industrial Chemist	X	X				X
Industrial Designer	X		X		X	
Industrial Engineer	X	X		X		X

	Realistic	Investigative	Artistic	Social	Enterprising	Conventional
Industrial-Organizational Psychologist	X	X	X			
Labor Relations Manager				X	X	
Manufacturing Engineer	X	X			X	X
Materials Engineer	X	X			X	X
Mechanical Drafter	X	X				X
Mechanical Engineer	X	X			X	X
Millwright	X	X				
Nanotechnologist	X	X				X
Pattern and Model Maker	X					X
Power Plant Operator	X					
Printing Press Operator	X					
Production Manager					X	X
Purchasing Agent				X	X	X
Quality Control Technician	X					X
Robotics Technologist						
Semiconductor Processor	X					
Sheet Metal Worker	X	X			X	X
Tool and Die Maker	X					X
Welder	X					X
Woodworker	X					X

Now it's time to move on to the next big step in the Big Question process. While the first step focused on you, the next one focuses on the world of work. It includes profiles of a wide variety of occupations related to manufacturing, a roundtable discussion with professionals working in these fields, and a mind-boggling list of other careers to consider when wanting to blend passion or talent in these areas with your life's work.

SECTION 2 explore your options

By now you probably have a fairly good understanding of the assets (some fully realized and perhaps others only partially developed) that you bring to your future career. You've defined key characteristics about yourself, identified special interests and strengths, examined your work values, and analyzed your basic work personality traits. All in all, you've taken a good, hard look at yourself and we're hoping that you're encouraged by all the potential you've discovered.

Now it's time to look at the world of work as it pertains to the manufacturing industry.

As you've probably learned in your studies about the Industrial Revolution, manufacturing has long played a central role in keeping our nation strong and prosperous. In recent years you may have caught some of the buzz about manufacturing jobs being outsourced overseas and may have even seen the closing of some manufacturing facilities in your hometown. So you may wonder why in the world we would be encouraging some of America's best and brightest young people to consider a career in manufacturing.

According to a survey conducted by the National Association of Manufacturing, a professional organization representing thousands of manufacturing companies, more than 80 percent of the survey manufacturers reported a moderate to serious shortage of qualified job applicants—"qualified applicants" being the operative words—even though manufacturing was suffering serious layoffs.

This factor, coupled with the trend of retiring baby boom generation workers leaving the workforce in droves over the next 15 to 20 years, will lead to a projected shortfall of 10 million new skilled workers by 2020. That makes for some interesting opportunities for your up-and-coming generation.

Of course, if you are like a lot of people, you may be unwittingly buying into some negative stereotypes associated with manufacturing.

fyi Each of the following profiles includes several common elements to help guide you through an effective career exploration process. For each career, you'll find

- A sidebar loaded with information you can use to find out more about the profession. Professional associations, pertinent reading materials, the lowdown on wages and suggested training requirements, and a list of typical types of employers are all included to give you a broader view of what the career is all about.
- An informative essay describing what the career involves.
- Get Started Now strategies you can use right now to get prepared, test the waters, and develop your skills.
- A Hire Yourself project providing realistic activities like those you would actually find on the job. Try these learning activities and find out what it's really like to be a . . . you name it.

You don't have to read the profiles in order. You may want to first browse through the career ideas that appear to be most interesting. Then check out the others—you never know what might interest you when you know more about it. As you read each profile, think about how well it matches up with what you learned about yourself in Section 1: **Discover You at Work.** Narrow down your options to a few careers and use the rating system described below to evaluate your interest levels.

- **No way!** There's not even a remote chance that this career is a good fit for me. (Since half of figuring out what you do want to do in life involves figuring out what you don't want to do, this is not a bad place to be.)
- **This is intriguing.** I want to learn more about it and look at similar careers as well. (The activities outlined in Section 3: **Experiment with Success** will be especially useful in this regard.)
- **This is it!** It's the career I've been looking for all my life and I want to go after it with all I've got. (Head straight to Section 3: **Experiment with Success.**)

When you think manufacturing you may think assembly line, sweatshop, or even worse. The truth of the matter, as you will see emphasized time and time again in the career profiles that follow, is that 21st-century manufacturing demands a new breed of worker: tech-savvy, highly skilled thinkers, doers, and creators.

Of course, exploring careers in any industry can be overwhelming, and the manufacturing industry is no exception. So here's another way (along with your Holland interest codes) to make it just a little easier to hone in on the types of careers most likely to be of interest to you.

Manufacturing careers can be grouped in the following six distinct categories. Understanding these pathways provides another important clue about which direction might be best for you. Following are more details about each of the six manufacturing pathways.

Production

Manufacturing is all about producing products. The production pathway includes all the careers involved in making, constructing, or assembling an enormous assortment of products. Production careers profiled in this book include assembler, cost estimator, electronic equipment assembler, foundry worker, hoist and winch operator, millwright, pattern and model maker, printing press operator, sheet metal worker, tool and die maker, welder, and woodworker.

Manufacturing Production Process Development

Manufacturing production process development includes those occupations involving the fulfillment of two important manufacturing distinctions: what products get made and how various products are made. On the "what" side of the equation are product designers and industrial engineers. On the "how" side of the equation are people who create the processes whereby products are made using technology, robotics, and all kinds of sophisticated machinery. Manufacturing production process development occupations profiled in this book include CAM technician, chemical engineer, computer hardware engineer, electrical engineer, industrial designer, industrial engineer, manufacturing engineer, materials engineer, mechanical drafter, mechanical engineer, nanotechnologist, robotics technologist, and semiconductor processor.

Maintenance, Installation, and Repair

Keeping equipment, machines, and tools in tip-top shape is the number one job priority of people working in maintenance, installation, and repair jobs. A big part of their work involves preventing problems, while another part involves solving problems as they occur in any situation involving humans, technology, and machinery. Among the maintenance, installation, and repair occupations profiled in this book are avionics technician, boilermaker, millwright, and power plant operator.

A Note on Websites

Websites tend to move around a bit. If you have trouble finding a particular site, use an Internet browser to search for a specific website or type of information.

Quality Assurance

Making sure that the production process and resulting products meet exacting standards and procedures describes the main function of quality assurance occupations. Various types of professionals are responsible for different phases of the production process. Some quality assurance professionals focus on the raw materials, others on manufacturing equipment, and others on the final product. You'll find a profile of a quality control technician in this book.

Logistics and Inventory Control

Movement is the key word associated with the logistics and inventory control pathway. Professionals with job titles such as dispatcher, material mover, logistical engineers, logistician, and traffic manager are responsible for getting raw materials, parts, and finished products where they need to be—whether it's to the production line or ultimately to the customer. Production manager and purchasing agent are logistics and inventory control professions profiled in this book.

Health, Safety, and Environmental Assurance

Safety is the number one priority for this pathway. This priority involves a variety of professions devoted to three key areas: people, products, and the environment. Tasks routinely performed within this pathway include conducting health, safety, and environmental investigations; establishing preventive measures; implementing health, safety, or environmental programs or policies; and planning for safety in new production processes. Environmental engineer, industrial chemist, industrial-organizational psychologist, and labor relations manager are professions associated with this pathway that are profiled in this book.

As you explore the individual careers in this book and others in this series, remember to keep what you've learned about yourself in mind. Consider each option in light of what you know about your interests, strengths, work values, and work personality.

Pay close attention to the job requirements. Does it require math aptitude? Good writing skills? Ability to take things apart and visualize how they go back together? If you don't have the necessary abilities (or don't have a strong desire to acquire them), you probably won't enjoy the job.

In the following section you'll find in-depth profiles of 35 careers representing the manufacturing industry. Some of these careers you may already know about, while others will present new ideas for your consideration. All are part of a dynamic and important segment of the U.S. economy.

assembler Is your room stocked with model airplanes you built yourself? Do you enjoy taking things apart and putting them back together? If you enjoy putting things together, can follow detailed instructions, and can work quickly and accurately, you might want to consider a career as an assembler or fabricator. Assemblers work for large companies that produce products (from toys to automobiles to space vehicles) or the components that make up those products (from engines to timers to machine switches and controls).

Other assets important for a career as an assembler include manual dexterity and good eyesight (with or without glasses) when working with small parts. Ever notice all the different colored wires in many electrical components or products? Plants that manufacture those items often test for colorblindness to be sure that workers can complete their tasks correctly.

Although some assemblers work on a product from start to finish, most assemblers work on teams. In some situations each assembler is highly trained in a specific task related to assembling, which, when everyone on the team does their part, ensures quality in the end product. However, many manufacturers have changed to a system of "cellular manufacturing," in which each member of the team (or "cell") rotates through all of the tasks assigned to that team, rather than specializing in a single task. Companies have found that workers tend to

Get Started Now!

- Visit an electronics, automobile, or other large manufacturing plant. Talk to the assemblers about their skills, working environment, and schedules.
- Take as many math and computer classes as possible.
- Play up your teamwork skills! Sign up with organizations that emphasize a team approach to making something, such as Habitat for Humanity (***www.habitat.org***).

Search It!
International Association of Machinists and Aerospace Workers at ***www.goiam.org***, Electronic Industries Alliance at ***www.eia.org***, and Electronics Technicians Association (ETA) at ***www.eta-i.org***

Read It!
Assembly magazine at ***www.assemblymag.com*** and ***www.4assembly.com/assyexp.html***

Learn It!
High school diploma, plus on-the-job training is required.

Earn It!
Median hourly earnings of team assemblers are $10.90. (Source: U.S. Department of Labor)

Find It!
See current job listings at the ETA website at ***www.eta-i.org/jobs.html***.

Hire Yourself!

Want to see how well suited you are for precision assembling? Origami, the Japanese art of paper folding, is a free and easy way to check out your manual dexterity and your ability to follow instructions.

You can get a library book on the subject or go to ***www.paperfolding.com***. Use the site to learn a little bit about the history and art of origami. Then click on "learn to fold" and see how well you can assemble at least three of the animals, flowers, and other designs. Use poster board or a decorated shoe box to display your work.

be happier and more productive when the process emphasizes teamwork and communication.

Working conditions for assemblers vary among industries, and ergonomic research has greatly improved workers' comfort levels. In some industries, such as electronic and electromechanical equipment assembly, work is done in rooms that are clean, well lit, and dust-free. Other manufacturing sectors—like aircraft assembly—require assemblers to get their hands dirty, especially when dealing with oil or grease components. But latex gloves and clean work suits help minimize the damage.

You may have heard about robots, technology, and computers reducing the need for assemblers. While advanced technology is a fact of life, there is still opportunity in this field. Older workers are retiring and need to be replaced, and there are many places where new technology is not usable. Small companies cannot justify the expense of this equipment. And some tasks, like working in hard-to-reach places in an airplane fuselage or a gearbox, are not suited to robots or other automation.

In addition, assemblers with math, science, and computer skills and an appreciation for how products are assembled are needed to operate and maintain the automated machines.

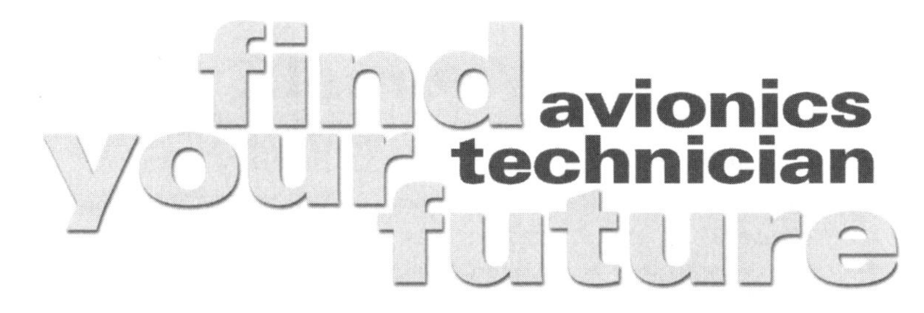

avionics technician

Have you ever thought about how much our country depends on airplanes? We need them to visit relatives in distant places, go on vacations, and check out faraway college campuses, but airplanes are critical to many other aspects of our society as well. We rely on airplanes to transport mail, to deliver organs for transplants, to bring emergency workers and supplies to areas hit by fires or hurricanes, to deliver parts for fast repair of important machinery, and even to get felons to the right prisons.

Avionics technicians are the well-trained professionals who keep airplanes in top working order, doing frequent checks and preventative maintenance before and after each and every flight. Many technicians specialize in one or more specific types of aircraft, such as jets, propeller-driven airplanes, or helicopters.

Avionics technicians are the people who repair and maintain all of the electronic components of the aircraft, including the instruments used for aircraft navigation and radio communications, the weather radar systems, and even the frequently heard about "black box" flight data recorder.

Sometimes avionics technicians specialize in a specific type of repair such as engines and propellers (a combination called the powerplant) or the airframe. However, most civilian aircraft mechanics are trained to work on all parts of the plane except instruments.

Get Started Now!

- Load up on math, physics, chemistry, and computer courses.
- If electronics or mechanical drawing courses are offered in your high school, be sure to enroll.
- Contact a small local airport to arrange to watch aircraft mechanics in action. Contact airlines to meet or interview one of their mechanics or avionics technicians.

Search It!
International Association of Machinists and Aerospace Workers at ***www.iamaw.org*** and Professional Aviation Maintenance Association (PAMA)at ***www.pama.org***

Read It!
Avionics magazine at ***www.defensedaily.com/reports/avionics*** and Aviation Week's AviationNow.com at ***www.aviationnow.com***

Learn It!
Training is obtained at an FAA-approved trade school. For a list of aviation training programs go to ***www.aviationschoolsonline.com***.

Earn It!
Median hourly wage is $20.21. (Source: U.S. Department of Labor)

Find It!
To see current job listings visit the PAMA website at ***www.pama.org*** and click on "Job Bank."

Hire Yourself!

Use the Internet to learn about what some of the different flight instruments do and how they work. A good place to start is at the Experimental Aircraft Association site at ***www.eaa838.org/instruments.htm***. Use poster board or a favorite computer program to create a diagram that identifies and describes the most common flight instruments.

To get an FAA (Federal Aviation Administration) aircraft mechanic certificate, the applicant must attend an FAA-approved school, have sufficient work experience, pass written and oral tests, and demonstrate proficiency in the desired specialty. Technicians without certificates can also work on planes, but must be supervised by those with certificates. Technically, on-the-job training can substitute for formal schooling, but that is becoming much less common.

Avionic technicians can advance to become authorized inspectors after they have held an A&P certificate for at least three years. Avionics technicians may also require additional licenses, such as a radiotelephone license issued by the U.S. Federal Communications Commission (FCC).

People who are successful in these careers are hard-working, self-motivated team players. They have a keen attention to detail, the ability to identify potential problems before they occur, and are good at working under pressure to meet strict deadlines.

Avionic technicians must be able to lift or pull objects weighing as much as 70 pounds, have high mechanical aptitude, be agile enough to work in hard-to-reach places, and comfortable working at great heights (on the tops of wings or fuselages). They must be able to diagnose mechanical problems from the pilot's description (unlike the mechanic who can take your car out and hear the weird noises for himself), and work in the atmosphere created by noisy, vibrating engines.

Avionics technicians need to be strong in electronics aptitude and able to understand and troubleshoot the complex instruments that are represented by the dozens of gauges, indicators, buttons, lights, and gadgets in the cockpit. And, because technology is always changing, technicians need to commit to ongoing training throughout their careers.

find your future

boilermaker

boilermaker If boilermakers and boiler mechanics didn't do what they do, we'd be living in a very cold world. People who work in the boiler industry manufacture the heat and power equipment that keep many homes and businesses at a comfortable temperature. Besides actually making the equipment, boilermakers also install it, maintain it, repair it, and replace it when necessary.

Boilers supply steam to drive huge turbines in electric power plants and to provide heat and power in buildings, factories, and ships. Tanks and vats are used to process and store chemicals, oil, beer, and hundreds of other products. Most of the vessels boilermakers build and maintain are dust, air, gas, steam, oil, water and other liquid-tight pressure vessels. Because most boilers last more than 35 years, boilermakers and mechanics constantly inspect boilers, using tools and the latest technology to keep equipment in peak shape.

Successful boilermakers need good manual dexterity, which is the ability to use one's hands to manipulate objects and handle the required tools. This career requires an affinity with a wide variety of tools including power hammers, levels, wedges, turnbuckles, wrenches, torches,

Get Started Now!

- Look through the local telephone book under "boilers" to find boilermakers, installers, or inspectors. Give them a call and ask for an informational interview to learn more about the field.
- If your high school or local community college offers courses in shop, welding, or metalworking, take them.
- Ask your school custodian or maintenance worker for a tour of the building's boiler room. It will give you a better idea of the type of equipment you'll be working on.

Search It!
American Boiler Manufacturing Association at ***www.abma.com*** and National Board of Boiler and Pressure Vessel Inspectors at ***www.nationalboard.org***

Read It!
Find out about different types of boilers at ***www.processregister.com***

Learn It!
- High school diploma plus apprenticeship, on-the-job training, and continuing education seminars and classes
- Apprenticeships can be found through the Boilermakers National Joint Apprenticeship Program (***www.bnap.com***)

Earn It!
Median hourly wage is $20.17. (Source: U.S. Department of Labor)

Find It!
Most boilermakers and boiler mechanics belong to a labor union, such as the International Brotherhood of Boilermakers (***www.boilermakers.org***).

Hire Yourself!

Imagine that you've just been accepted into a boilermaker apprenticeship. You want to make a good first impression by being well informed about the types of tools boilermakers use in their work. Use information like that found at the How Stuff Works website at ***www.travel.howstuffworks.com/steam2.htm*** and the Aalborg Industries website at ***www.aalborg-industries.com/ifs/files/AI/eng/Presentation/Website/Marine/Manufacturing/process.jsp*** to put together a "virtual" toolbox. You can use a favorite Internet search engine such as Google or Yahoo! to gather additional information. Your toolbox should include pictures and descriptions of a plumb bob, level, wedge, turnbuckle, valve, acetylene torch, jack, caulking hammer, threading die, power saw, file, power grinder, and other tools you may learn about through your research.

saws, welding equipment, and other hand and power tools to get the job done. Physical strength is a required attribute as well—many of those tools are fairly heavy.

Small boilers can be assembled at the manufacturing plant, but most boilers are huge cylindrical containers that must withstand serious pressure from the liquids and gases stored within them. Because of the size and weight, they are made in sections out of iron or steel, and put together on site. Boilermakers follow blueprints to figure out how to install the boiler onto its foundation and decide which rigging materials are necessary to hoist the parts into place. At the boiler site, they install the boiler and bolt or weld the equipment. Next, they attach and align valves, tubes, gauges, and other necessary parts. Finally, they check for faults or leaks. Each process requires painstaking precision and accuracy.

The best way to learn this trade is through a four-year formal apprenticeship that combines on-the-job training with 144 hours of classroom instruction per year. This course of study covers almost everything you'll need to know to start up in this field: set-up and assembly, welding, blueprint reading, laws, operation, maintenance, and safety. The education doesn't stop there. Experienced boilermakers keep current by attending classes and seminars offered through the American Boiler Manufacturing Association and the National Board of Boiler and Pressure Vessel Inspectors.

Another option in the field of boilers is that of boiler inspector. This professional inspects boilers, enforces safety standards, and relays information to the public, boiler owners, and boiler manufacturers. Boiler inspectors investigate violations, accidents, explosions, and complaints—and make sure they rarely happen. While most inspectors have their high school diploma and on-the-job training, some boiler inspectors obtain an associate's degree in mechanical engineering. This additional educational credential often gives them an extra edge at promotion time.

Boilermakers often use potentially dangerous equipment, such as acetylene torches and power grinders, handle heavy parts, and work on ladders or on top of large vessels. Work may be done in cramped quarters inside boilers, vats, or tanks that are often damp and poorly ventilated. To reduce the chance of injuries, boilermakers may wear hardhats, harnesses, protective clothing, safety glasses and shoes, and respirators. Boilermakers work in boiler manufacturing shops, iron and steel plants, chemical plants, shipyards, and other types of large industrial and commercial complexes.

Search It!
Computer Applications for Manufacturing Users Society (CAMUS) at ***www.camus.org***

Read It!
CadCam magazine at ***www.delcam.com/info/latest_issue.htm***, *Computer-Aided Engineering* magazine at ***www.caenet.com***, and *Cadalyst* magazine at ***www.cadalyst.com***

Learn It!
- Two-year degree from a technical school or community college (click on "certified schools" at ***www.adda.org***)
- Certification from an AutoCAD training center (see ***http://usa.autodesk.com***)

Earn It!
Median annual salary is $41,170. (Source: U.S. Bureau of Labor Statistics)

Find It!
See job postings at CAD Job Mart (***www.cadjobmart.com***) and CAMUS (***www.camus.org***).

CAM (computer-aided manufacturing) technician

Before the newest "concept cars" make it to the auto shows, before they exist even as tiny models, they are first designed and tested in cyberspace. Thanks to the wonders of computer-aided engineering, these and countless other manufactured products are now created through a streamlined and very effective process that begins in the mind of an engineer and ends in a manufacturing production facility.

An engineer's ideas can be designed, analyzed, and produced with more creativity and accuracy using sophisticated software programs that include computer-aided design (CAD) technology. CAD technology helps designers visualize virtually any conceivable product in three-dimensional (3-D) format. This breakthrough technology allows engineers and other professionals to produce complex virtual models of

Get Started Now!

- Take as much math as possible, especially geometry and algebra.
- Learn your way around computers by taking a variety of courses in computer science and computer graphics.
- Check out the award-winning entries for the CADDIES, *Cadalyst* magazine's award program for CAD design, at ***http://aec.cadalyst.com/aec/article/articleDetail.jsp?id=104693***. You'll find designs for everything from airport terminals to government centers to museums.

Hire Yourself!

Go on-line to find out how pencils are made. Good starting points for your research include the General Pencil website (***www.generalpencil.com/how.html***), the Musgrave Pencil Company website (***www.pencils.net/slats.cfm***), and the How Stuff Works website (***http://science.howstuffworks.com/question465.htm***). Use what you learn to map out an efficient step-by-step manufacturing plan. Depending on your expertise and available resources, you can use either pen and paper or a computer drawing program to sketch out your plan.

almost anything, using true-scale measurements. Thus, an engineer working on an automobile could use CAD and other types of special programs to test the aerodynamics, handling, and interior comfort of the car—all on the computer screen.

From cars to machine parts to buildings, CAD (sometimes referred to as CADD, computer-aided design and drafting) has revolutionized the world of drafting. Where in the past detailed diagrams were drawn at a drafting table using T-squares, now CAD technicians input lines, arcs, coordinates, and dimensions into a computer. Once a model is drawn, CAD technicians can easily explore a greater number of design alternatives—reducing design costs and the time it takes to deliver a product to market. While trained engineers and designers usually create the concepts for the product, many of the actual drawings and adjustments are done by CAD technicians.

Once all the designs are complete and the product is ready to be produced, professionals once again rely on computers to help plan and implement the manufacturing processes using computer-aided manufacturing (CAM) technology. CAM processes automate manufacturing systems by sending detailed and very precise work instructions to various types of manufacturing machinery. For instance, a computer may be programmed to control an entire manufacturing process whereby raw materials may be moved from a robotic milling machine to a lathe, over to a welding machine, and then on to other tools until the intended product is completely formed.

Smart as it is, a CAM system cannot generate these complex instructions on its own. That's where the CAM technician comes in. He or she must program the system with coded instructions that make the machines perform the required tasks in the requisite order. Technology helps simplify this enormously complicated process. Once the technician knows

how to use CAM technology, generating instructions—and even accommodating last minute changes—can be accomplished through a few well-educated keystrokes.

Increasing numbers of community colleges and vocational/technical training programs offer programs in CAD, CAM, and other types of manufacturing. Students entering either a two-year associate's degree or a technical certification program can expect to take courses that challenge them to use algebra, geometry, communications, electronics, physics, and drafting in a manufacturing environment. They'll gain an in-depth understanding of how and why various manufacturing systems do what they do, and they should get some hands-on experience using a variety of manufacturing equipment. Graduates of these types of programs are prepared to launch a manufacturing career with companies that produce or process electronic systems and components, precision machine parts, wood products, plastic products, food products, and countless other products used by consumers and industries around the world.

chemical engineer

Chemical engineers are neither engineers who make chemicals, nor chemists who build things. All engineers are basically problem solvers. Chemical engineers solve problems that have to do with chemicals, including their properties, their reactions, and their uses. While chemists discover new chemical compounds and understand how those compounds behave, chemical engineers find and implement efficient applications of those compounds.

Chemical engineers develop ways to manufacture, transport, and store both the compounds created by the chemists and the products that can be made with those compounds. They have a part in the manufacturing of almost every product we use, including food, packaging, plastics, pharmaceuticals, film, fabrics, paint, and petroleum. They are critical to environmental safety and cleanup, and to health and safety in manufacturing plants.

Chemical engineers may choose to specialize in an industry producing a specific product. They may choose to specialize in a specific operation, such as oxidation or polymerization, or they may choose to focus on a broader issue, like pollution control or chemical transportation and storage.

Get Started Now!

- Load up on science, math, physics, and chemistry courses.
- Take a look at some of the student resources and information available at the American Chemical Society website at ***www.chemistry.org/portal/a/c/s/1/educatorsandstudents.html***.
- Check the labels on everything you use, from foods to detergent. Get acquainted with the scientific names of chemicals and the products in which they're used.

Search It!
The History of Chemical Engineering at ***www.pafko.com/history*** and the American Chemical Society at ***www.acs.org***

Read It!
Chemical & Engineering News at ***www.pubs.acs.org/cen***

Learn It!
- Bachelor's degree in chemical engineering
- Master's or Ph.D. degree preferred

Earn It!
Median annual salary is $72,490. (Source: U.S. Department of Labor)

Find It!
Listings of current chemical engineering jobs can be found at the American Institute of Chemical Engineers website (***http://develop.aiche.org/careerengineer***).

Hire Yourself!

Congratulations! A major chemical company has hired you to help it define areas in which to look for new chemical developments. Look around your home, school, or community for ideas of where new chemicals could be used to improve the quality of life. Here are examples of things that have already been done (not that you couldn't refine them further!):

- New paints for preschools and children's bedrooms include chemicals that make crayon marks wash right off.
- A new "second skin" for soldiers can be sprayed or rolled over their injuries, protecting them until they reach medical care.
- New fibers for children's sleepwear are soft to the touch, but will not burst into flames or melt onto the child's skin when hit by a spark.

Look at the furniture and appliances in your home, the equipment emergency workers use, and the equipment you use for your favorite sports. Define the properties for the chemical discovery you would like to see, and how it would affect products made from the new chemical. Summarize your findings in a chart or with written (and very descriptive) recommendations.

Chemical engineers need to be "big picture" people. They have to be able to think about all of the impacts of a new product or process, from the safety of the workers producing it all the way through to the safety of the ultimate consumers. They need to think about the impact of the manufacturing process on the environment as well as the impact of using and ultimately disposing of the final product and the byproducts of its manufacture. They must be analytical, alert to every detail, and focused on finding workable solutions when they are not immediately obvious. Chemical engineers should also be organized and able to plan and prioritize their work. Since computers play a large part in research, production, analysis, and quality control, chemical engineers must know their way around some rather complex technology as well.

Chemical engineers can work in many different situations. Working with new, experimental, and potentially dangerous chemicals requires strict adherence to safety standards and the use of gas masks, safety goggles, and laboratory suits. They may have to work overtime to meet deadlines, and they need to be able to perform under the stress of the unforeseen problems that inevitably arise.

Chemical engineers have helped in the development of many products that help us in our daily lives. Their "big picture" view has helped make many chemical discoveries workable. For example: the discovery of chemicals that could fertilize plants to produce larger and more hardy food or flowers had to be balanced with what that fertilizer did to the people who ultimately ate the food, how it affected the long range use of the land on which the food or flowers were grown, and whether it impacted local wildlife or insects.

Over time chemical engineers find a variety of career options available to them. Many go on to work in marketing and sales in their field, and quite a few have become CEOs of large companies. Although the preparation for this field can be demanding, the ultimate payoff can be quite generous. For instance, a 2001 survey by the National Association of Colleges and Employers found that bachelor's degree candidates in chemical engineering received starting offers averaging $51,073 a year, master's degree candidates averaged $57,221, and Ph.D. candidates averaged $75,521.

Search It!
Institute of Electrical and Electronics Engineers at ***www.ieee.org***

Read It!
Machine Design at ***www.machinedesign.com***, EDN Access at ***www.e-insite.net/ednmag***, and "The Red Hill Guide to Computer Hardware" at ***www.redhill.net.au/ig.html***

Learn It!
- B.S. degree in computer science for entry-level jobs
- Master's degree or Ph.D. are becoming required by more employers

Earn It!
Median annual salary is $72,200. (Source: U.S. Department of Labor)

Find It!
Computer hardware designers work for companies that manufacture computers and their components such as IBM (***www.ibm.com***), Apple (***www.apple.com***), and Intel (***www.intel.com***).

computer hardware engineer

Hardware engineers or designers are the people who make computers possible. Hardware (the computer chips, circuit boards, keyboards, monitors, mice, printers, and everything else that goes with them) is the essence of computing. While software engineers design the systems that enable computers to do all the amazing things we do with them, hardware engineers provide the tangible technology that makes up any given computer system.

Hardware engineers are constantly improving the computers and accessories that are available. You may have noticed the results of their fast-faced progress when your new state-of-the-art computer seems obsolete almost overnight. Years ago, software engineers wrote complex programs that may have taken hours or even days to run. Today's hardware can run many of those same programs in just minutes, or even seconds.

Get Started Now!

- Get acquainted with your computer and other tech gadgets. Knowing what computers are made of and how they and their components work is the first step in this career. Check out ***www.howstuffworks.com*** for information on personal computers, parallel and USB ports, and different types of computer memory (RAM, ROM, etc.).
- If your high school offers advanced computer courses, add them to your schedule. If not, look into programs offered at computer centers or local community colleges.
- Take communications classes to hone your writing and presentation skills. Computer hardware engineers have to effectively communicate their ideas to peers.

Hire Yourself!

Give your computer a makeover! As a computer hardware engineer, your job is to make computers more attractive, more efficient, and more user-friendly. Design a computer—or a piece of computer hardware, such as a monitor—that meets all three of the above requirements. Draw a picture of your design—complete with labels—and present your idea to a group. What were their opinions or suggestions for your design?

Hardware engineers work on improving and perfecting desktop and notebook computers, giant servers, and video game equipment that we usually associate with the term "computers." But the field of embedded computer technology (where computers are used to run other devices) includes the computer chips that help cars to run; allow cell phones, coffeemakers, and dishwashers to work; and make talking dolls and robotic dogs possible.

In addition to designing, developing, and testing computer hardware, hardware engineers often supervise the manufacture and installation of the hardware as well. They design methods for various components to interface, or work together, including the interaction between central processing units (CPUs) and peripheral units (hardware) and between the operating system and user programs (software). They also design ways to make computer use more efficient. For example, a computer that's more than five years old has lots of wires hooked up to the back of the hard drive, with each wire connecting to a different piece of hardware, such as a printer, a fax, a modem, or the monitor. Thanks to hardware engineers, many of those wires have been replaced by ports that can handle a bunch of hardware with just a few plugs.

Hardware engineers help companies analyze their computing needs and recommend hardware that will upgrade or replace the current equipment. They may also train people to use the new system, and monitor its performance and maintain it.

Hardware engineers are very similar to electrical and electronics engineers, except that they work only with computers and computer-related equipment. Few academic programs are specifically designated for hardware engineers, so most have majored in electrical or electronics engineering or computer science. Some hardware engineers also have degrees in math or physics. Licensing is required in many states, and ongoing education to keep up with advances in the field is required.

Search It!
Society of Cost Estimating and Analysis at ***www.sceaonline.net***, American Society of Professional Estimators at ***www.aspenational.com***, and Institute for Supply Management at ***www.ism.ws***

Read It!
"But What Will It Cost?" at ***www.jsc.nasa.gov/bu2/hamaker.html*** and *ASPE Technical Journal* at ***www.aspenational.com/1technicaljournal.html***

Learn It!
Minimum of a four-year degree in engineering, operations research, mathematics, statistics, accounting, finance, business, economics, or a related subject.

Earn It!
Median annual salary is $47,550. (Source: U.S. Department of Labor)

Find It!
Find information about opportunities for cost estimators at ***www.aspenational.com/1opportunities.html***.

cost estimator

What does it take to make a company a financial success? Of course, the product has to be what customers want, of high quality, and easy to find and order. Service to customers has to be fast and efficient. But while excellent products and service may generate lots of sales, it is the accuracy of the cost estimators that ultimately ensures a profitable bottom line.

Whether a company is selling yachts or yo-yos, it can only make a profit if a product's selling price is more than the total of all of the costs. The same goes for companies that produce the raw materials (like steel, paper, or plastic), components (tires, transmissions, or radios), or machinery for other manufacturing companies.

The first thing the cost estimator has to do is determine what will go into the total cost of the product. That has to include the raw materials; the cost of any special machinery that may have to be modified, purchased, or rented; any necessary computer hardware and software; and the cost of labor (paying the people who do the work).

Cost estimators spend a lot of time with the engineers and designers to make sure that they really understand the product and what materials will be needed. For new products, it is important to be able to read blueprints to determine what tools, gauges, screws, and nuts and bolts will

Get Started Now!

- Learn as much as you can about the manufacturing process of the industries you might be interested in. Courses in drafting and blueprint reading will be a definite help.
- Take as many math courses as possible!
- Get comfortable using computer programs for spreadsheets and word processing. Business law and basic accounting will also help you become more familiar with some of the things cost estimators have to know.

Hire Yourself!

You have just been hired as the cost estimator for Best Careers Press, a major publishing company. They want you to estimate the cost of printing 100,000 copies of *Career Ideas for Teens in Manufacturing*.

First, determine the quantity of raw materials you'll need. In this case, that includes paper for interior pages, cardstock for the covers, packaging materials, and lots of ink! Search the Internet to find at least three vendors who can bid on each of the following: paper, printing services, book binding services, and packing materials. You may want to compare prices of full-service companies that can do it all and those that specialize in a specific product or service.

be needed. Once the estimators have a list of everything that is required, they find suppliers who can provide costs for those materials.

To calculate the cost of labor, estimators calculate the time that will be required to set up all of the production equipment, and then the time to produce each unit. The number of people hours is then multiplied by a standard wage rate to get the cost of labor. Estimators include more time for the beginning of the production process, when people are just learning their jobs. They also include extra time and materials for waste and practice or test runs. Cost estimators must have excellent math and computer skills to do all of these calculations.

Estimators are sometimes called upon to help set a profitable selling price for a product. They can be a critical part of the new product development process. If the cost estimator calculates that the cost of producing an item will be more than anyone will be willing to pay, the item will have to be redesigned, or maybe not produced at all. Another important aspect of the job is tracking the actual costs as a project proceeds, ensuring that they stay in line with the estimates.

Good cost estimators must have excellent interpersonal skills to work with engineers, designers, and clients, as well as with production workers. They have to be able to locate and negotiate with different suppliers to get the best costs and delivery times for the materials and components they need.

Estimating can be an exciting and rewarding career for someone who enjoys working with numbers. Estimators usually work in corporate (or government) offices, but also get to visit production facilities. They have the satisfaction of knowing that they are helping the company produce high quality products at the lowest possible cost and highest profit.

Search It!

Institute of Electrical & Electronics Engineers at (IEEE) ***www.ieee.org***

Read It!

EC&M (*Electrical Construction & Maintenance*) at ***www.ecmweb.com***

Learn It!

- B.S. degree in electrical engineering for entry-level jobs
- M.S. degree or Ph.D. for advanced-level jobs

Earn It!

Median annual salary is $68,180. (Source: U.S. Department of Labor)

Find It!

Electrical engineers are in demand at manufacturing companies, technology companies, communications corporations, and power companies. Many work in the communications, defense, and aviation industries. Find information about a wide variety of opportunities at the IEEE Job Site (***www.ieee.org***).

electrical engineer

Are you fascinated by electronics gear? Interested in all the details of what distinguishes one stereo system from another? Excited about each new generation of cell phones with cameras and other bells and whistles? Is "technology" your preferred language? If so, you may find yourself feeling right at home among electrical or electronics engineers.

Electrical and electronics engineers design, develop, and test electrical equipment and components. They may also supervise the manufacturing and installation of electrical systems for commercial, industrial, military, or scientific use.

Although the terms "electrical engineer" and "electronics engineer" are often used interchangeably, some schools and job listings do differentiate between them. In general, electronics engineers deal with communications, home appliances, and other low power level projects. Electrical engineers are more likely to work with the power grids of utility systems, with industrial power systems, and with other large power situations. Of course, there is a lot of overlap between the areas.

Like all engineers, electrical and electronics engineers are basically problem solvers. Most tend to specialize in one of three areas: commu-

Get Started Now!

- Take as many math courses as you can, including calculus and statistics if they are offered, as well as business math such as accounting.
- Get involved with your school's science club, math group, computer club, or physics team.
- Talk to an electrical engineer—better yet, a bunch of them—about the type of work he or she performs. Try to find engineers who work for different sectors so you can compare the job responsibilities and requirements.

Hire Yourself!

Electrical engineers are constantly working to improve technology and products—but they never get due credit because much of their work is done behind the scenes. Use the Internet or recent news magazines to learn about a new tech gadget that owes its thanks to an electrical engineer. Write up a press release about the product, the engineer, and how the technology will impact people or the future. If you can, interview the engineer or company responsible for the product. Submit the press release to your school or local newspaper to help spread the word.

nications, electrical equipment, or power generation, transmission, and distribution.

- Communications projects include everything from cell phones to wireless Internet connections, high-speed modems, and satellite transmissions. Communications today can be transmitted by wire, microwave, or fiber optics. Signal processing (coding the electronic signals for transmission, and then decoding them so that they can be understood as words, music, or images) is an important part of the communications field.
- Electrical equipment engineering includes designing, testing, and manufacturing all types of electronic equipment, from aviation controls to robots, toys, and even slot machines. Work in this area focuses on understanding circuits, and creating products that work more effectively and can be produced less expensively. Although computers seem to fall into this area, engineers who work with computers are called computer or hardware engineers.
- Power engineering involves creating energy, or power, and getting it from one point to another. Understanding electromagnetics (magnetic fields caused by an electric current) is a big part of this field. Electrical engineers need to consider the effects of the electromagnetic field caused by the current in energized power lines, and how that field can interfere with nearby data cables.

Important skills for electrical/electronics engineers involve mastery of high-level math and science concepts. Algebra, geometry, calculus, statistics, and physics are all used on a daily basis. Computer skills are essential, and computer-aided design (CAD) systems are regularly used to create realistic models that can simulate the engineer's designs and find the potential benefits and shortfalls. Good communication skills also come in handy since electrical and electronics engineers have to explain their work to clients and management, and write clear reports on what they have done or would like to do. Patience is another big asset. Even if an engineer creates the best gadget ever to see the light of day, the average span of time from the design of a product to placement on a shelf is still two years.

Electrical and electronics engineers have good starting salaries. According to a 2001 salary survey by the National Association of Colleges and Employers, bachelor's degree candidates in electrical and electronics engineering received starting offers averaging $51,910 a year; master's degree candidates averaged $63,812; and Ph.D. candidates averaged $79,241.

To continue to grow in the field, most electrical/electronics engineers pursue advanced degrees and get licensed by their state. Getting licensed requires four years of work experience plus a state exam before being certified by the Accreditation Board for Engineering and Technology.

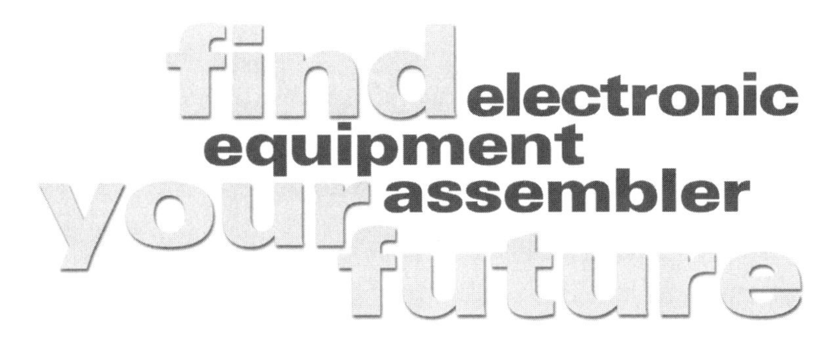

electronic equipment assembler

Have you ever fixed a camera, computer, TV, CD player, computer game pad, or any other gadget? Then you can appreciate the job that electronic assemblers do. According to the U.S. Department of Labor's *Occupational Outlook Handbook*, electrical and electronic equipment assemblers put together or modify missile control systems, radio or test equipment, computers, machine-tool numerical controls, radar or sonar systems, and prototypes of these and other products. In simple terms, their job is to put together the wiring and inner "guts" of a product to make it work.

This career requires more than just tinkering with wires and tools. It's a hands-on job that requires good eyesight, manual dexterity, and the ability to conduct complex, repetitive tasks quickly and methodically. The job also requires keen thinking skills coupled with an ability to read and interpret engineering specifications from text, drawings, and computer-aided drafting systems. Assemblers also may use a variety of tools and precision measuring instruments. Some experienced

Get Started Now!

- If your school offers shop, welding, or electronics classes, by all means add them to your schedule. It'll give you an edge over other entry-level assemblers.
- Look for summer work, work-study, or internship opportunities in a retail store that specializes in electronic products.
- Hone your reading, writing, and verbal skills. Assemblers who can understand detailed directions and instruct others on how to work have better chances of advancement.

Search It!
The Electronic Industries Alliance at ***www.eia.org***

Read It!
Circuits Assembly at ***www.circuitsassembly.com*** and *Semiconductor International* at ***www.semiconductor.net/packaging***

Learn It!
High school diploma (or equivalent) plus on-the-job training is required.

Earn It!
Average annual salary is $23,200. (Source: U.S. Department of Labor)

Find It!
Electronic equipment assemblers work for large manufacturing companies such as Intel (***www.intel.com***) and Panasonic (***www.panasonic.com***).

Hire Yourself!

Go to the "electronics stuff" section of the How Stuff Works website (***http://electronics.howstuffworks.com***). Choose an electronic product that interests you and write a set of instructions about how to assemble it. Use words and diagrams to clearly explain each step of the process.

assemblers work with engineers and technicians, assembling prototypes or test products.

Some electronic components are so small that assemblers need microscopes to do their jobs. Others require the use of hand tools, power tools, and welding equipment to put parts together. Some specialized assemblers also use cables, tubing, and circuits to do their jobs.

Having the ability to work with a team as well as independently is key. In some companies, assemblers work on a production line, and each person has a specific task, such as pressing components into place. In other companies, assemblers perform several different tasks or may create a product from start to finish. Still other companies use assemblers to monitor the machines that do the repetitive work.

As technology continues to change the manufacturing process, versatility has become a critical credential for electronic equipment assemblers. Technologies such as robotics, computers, programmable motion control, and various sensing technologies change the way in which goods are made and affect the jobs of those who make them. The ability to maintain and repair this type of automated equipment provides other, more advanced employment opportunities.

Other ways to stay marketable in this field involve acquiring training in specialties such as electronic technology or advanced instruction in soldering and assembly. Assemblers who round out their résumés with training and experience can move on to become electronics testers, inspectors, or electronics technicians.

environmental engineer

When the United States first began to be the major source of manufactured goods, industry's biggest concerns were getting out quality products at a reasonable cost. America's successful manufacturing growth created the highest standard of living the world has ever known. As a result the United States became a place where everyday people could have cars, washing machines, cosmetics, drugs, and computers.

At that time, no one thought much about how all this manufacturing affected the environment or the workers in the plants. Eventually experts realized that waste products created in the manufacturing process were polluting the soil, air, and water supplies; that workers in these plants often developed life-threatening illnesses; that the supply of raw materials used in manufacturing was not infinite; and that there was a limit to where waste materials could be disposed.

These issues gave rise to a new scientific specialization and ushered in a new breed of engineer—the environmental engineer. Environmental engineers take a long, hard look at how human activity impacts the environment—and what can be done to minimize the damage. Environmental engineers are often the first on the scene when an oil spill or other major catastrophe damages the environment or harms wildlife, measuring the pollutions and supervising cleanup efforts.

In the manufacturing sector, environmental engineers use the principles of engineering, biology, and chemistry to prevent and repair damage to the environment or to assess environmental health hazards caused by the manufacturing process.

As part of their job, environmental engineers may design systems, processes, and equipment to achieve these goals. They develop long-term plans for manufacturing facilities to change their methods and materials to comply with government regulations, and then monitor and report on the implementation of these plans. They develop health and

Search It!
American Academy of Environmental Engineers at ***www.aaee.net***

Read It!
Journal of Environmental Engineering at ***www.pubs.asce.org/journals/ee.html***

Learn It!
Bachelor's degree in engineering with an emphasis on environmental studies is typical.

Earn It!
Median annual salary is $61,410. (Source: U.S. Department of Labor)

Find It!
Environmental engineers find work in the chemical and petrochemical industries as well as power generation. They may also work with natural resource industries such as mining, oil, and gas, in waste management companies, and for the federal government. Search the Environmental Protection Agency website at ***www.epa.gov*** for more information.

Get Started Now!

- Think globally, act locally! Start or join an environmental group in your high school or your community. Become involved in environmental projects—besides looking great on your college application, you'll be benefiting the planet!
- Learn about cases in which environmental engineers made a real difference. Check out *Erin Brockovich*, a movie based on a real person who put together more than 600 people to go after PG&E, a $30 billion company, which polluted the environment in its area. Study the Love Canal case, in which Hooker Chemical dumped toxic chemicals into the local water, causing extremely high rates of cancer and birth defects. (A good reference on Love Canal is at ***www.epa.gov/history/topics/lovecanal/01.htm***.)
- Check out Partnership Projects—Design for the Environment, US EPA (at ***www.epa.gov/opptintr/dfe/projects/index.htm***) for an overview of approaches and tools for environmental management and for the specific needs of 13 industries.
- Improve your writing and communication skills. Writing reports, conducting research, and explaining the environmental impact of a situation is all in a day's work.
- Take classes in biology, chemistry, and other sciences.

safety protocols specific to each manufacturing site, developing and publicizing plans for contingencies like spills or accidents, and methods for loading and transporting waste. Many environmental scientists work for government agencies, advising on the need for new regulations or inspecting facilities to be sure that they meet existing regulations.

In recent years, there has been a shift in emphasis toward preventing problems rather than controlling those that already exist. As a result, in addition to their involvement in the actual manufacturing process, environmental engineers also help to manage the natural resources (such as forests and agricultural land) from which products are made, find ways to use recycled materials and to use materials that can be recycled in the future, and ensure that all products are safe for the ultimate consumer.

The U.S. Department of Labor forecasts great growth in environmental engineering jobs. More concern about the environment, coupled with increased government regulation and local concern, means that environmental engineers will be called upon in more and more situations.

College courses include high-level math and physical sciences, engineering, and classes that concentrate on the environment. After obtaining a bachelor's degree in engineering, environmental engineers take courses in an accredited program and must pass a state examination. Most—if not all—environmental engineers agree that continuing education is also a must. Changes in technology and the environment help shape and change an environmental engineer's role constantly.

Hire Yourself!

Environmental engineers have to be up to speed on the latest laws and regulations affecting our environment. Search the Internet or news magazines for recent laws that have had an impact on the environment—positive or negative. A good place to start your research is at the Sierra Club website (***www.sierraclub.org***). Brainstorm. What are some ways you can use this law to make a change in your community or local companies? Share your recommendations in a briefing paper or a PowerPoint presentation.

Search It!

American Foundry Society at ***www.afsinc.org***, Castings Technology International at ***www.castingsdev.com***, and Foundry Educational Foundation at ***www.fefoffice.org***

Read It!

Modern Casting magazine at ***www.moderncasting.com***

Learn It!

- High school diploma
- On-the-job training and an apprenticeship that may last up to five years

Earn It!

Average annual salary is $28,000. (Source: U.S. Department of Labor)

Find It!

Foundry workers are employed in independent foundries and in manufacturing companies, usually through large labor unions such as the United Steel Workers of America (***www.uswa.org***).

foundry worker

Take a look around you. Do you see any cars? Machines or products with metal components? Tools, wheels, or machines made of metal? In fact, most things you see probably have some metal parts or were made with machines that used metal parts. Who makes those metal parts? Foundry workers specialize in manufacturing metal parts made of steel, iron, bronze, brass, or aluminum.

A typical day on the job involves molten metal, powerful machinery, and heavy molds and cores. The work requires a variety of tasks that are generally performed by workers trained in flawlessly executing each procedure. The following are some of the skilled workers involved in making metal castings:

- Patternmakers use a variety of tools and machines to make a full-scale wooden or metal model of an object according to engineering specifications

Get Started Now!

- Take shop or metalworking classes if they're offered in your high school. If none are available, search your local phone book for metal, glass, or wood studios and ask about private or group classes.
- Learn about the sand used in foundry work and how some of it is recycled from Foundry Industry Recycling Starts Today (FIRST) at ***www.foundryrecycling.org***.
- Sometimes it's easier to understand foundry work and the casting process by looking at smaller projects. The website for the Le Blanc Foundry (***www.leblancfineart.com/foundry/casting/stage1.htm***) shows the steps in one form of sculpture casting.

Hire Yourself!

China has become one of the world's biggest consumers of steel. You're the director of new business development at an American steel company. Use the Internet to research this business trend and come up with five strategies for tapping this new market.

- Molders make a mold of the design using mixtures of sand and binding agents and open-ended containers called flasks. Molds are then used almost like "cookie cutters" to create multiple copies of the exact same pattern.
- Coremakers are called into action when a casting is meant to be hollow. Once again relying on a mixture of sand and binding agents, they create a core from the mold.
- Grinders finish castings by removing excess metal using tools that include pneumatic hammers and power abrasive wheels.

Other workers employed by foundries include those who operate the furnaces needed to melt metal or fire castings, those who apply finishes to the castings, those who pack castings for transpiration, and those who handle the day-to-day operations of the factory or plant.

Quality is a big issue. Foundry workers are responsible for ensuring that the metal and cast are up to standards—and that the finished product looks exactly like it's supposed to and has the strength it's intended to have. If only for this aspect of the process alone, this line of work requires skill and sound judgment.

Why the term "foundry"? We usually think of "found" as having located something that was lost, or as starting something (as in founding a local chapter of a national organization). But a third definition of "found" is based on the Latin "fundere" and means to melt and shape metal using castings or molds. And that's exactly what foundry workers do, making metal castings for missile parts, sewer pipes, stove doors, and more.

Like many jobs based in factories, foundry workers have to put up with some dust, heat, and quite a bit of noise from the machines and tools used to cut wood and metal. At times, the work requires physical brawn, especially when large metal castings need to be lifted and hauled.

Some foundry workers use their skills to create some of the world's most impressive symbols of beauty and artistic expression. Talk a walk around any community and you're likely to find statues or other sculpture made of bronze or other metals. These sculptures were initially

created in clay or some other substance and then cast at a foundry into metal. Some of our most famous national symbols were produced in this way. (Hint: Two of the best known have "Liberty" in their names.)

Because factories are quickly becoming automated, it's a good idea for prospective foundry workers to have some technical knowledge. Understanding how to use and maintain computerized machines can benefit workers who are starting out in the field—or looking for advancement. Entry-level foundry workers may go on to become supervisors, testers, or managers—especially when their experience is coupled with the people skills associated with effectively managing projects and other workers.

hoist and winch operator

Hoist and winch operators have quite a heavy load to carry. That's because a person who works as a hoist and winch operator lifts, pulls, and hoists hefty loads with equipment. Hoists and winches are specialized equipment used to lift, pull, and move heavy loads. Skilled operators are required to run and maintain that equipment.

Hoist and winch operators do much more than just attach hooks and lift or pull things. They are the experts who evaluate the size, shape, and weight of the materials and decide how they should be used. Their decisions about whether to use hoists or winches, platforms, metal cages, or hooks and cables are based on their knowledge of the capabilities of each type of machine and their own experience. There is no reference chart with all the answers. Instead, the job requires skill, know-how, and common sense. Sometimes operators need to use a truck or car to move

Get Started Now!

- Visit a large construction site or manufacturing warehouse and ask to speak to a hoist and winch operator. Learn about the work conditions and job hours, and ask to watch the operator in action.
- Take shop or mechanical classes while in high school. They'll help you build manual dexterity and improve hand-eye coordination.
- Look for summer jobs or internships in construction or farming that will give you experience on any type of heavy machinery. Actually operating machines can provide invaluable experience.

Search It!
International Union of Operating Engineers at ***www.iuoe.org***

Read It!
Learn about hoists and winches at ***www.littlehercules.com*** and about how hoists are sued in elevators at ***http://science.howstuffworks.com/elevators4.htm***

Learn It!
High school diploma plus apprenticeship or on-the-job training is typical.

Earn It!
Median hourly wage is $15.09. (Source: U.S. Department of Labor)

Find It!
Hoist and winch operators work for manufacturing companies, heavy construction companies, sawmills, planning mills, granite companies, and equipment rental companies.

Hire Yourself!

What's the difference between a hoist and a winch? A winch is for pulling; a hoist is for lifting. Use this newfound information and an Internet search engine to create a catalog of hoists and winches used in a variety of situations. Print out pictures and descriptions of each product and compile them in a four-page sales catalog.

the materials. They use jacks, slings, cables, and stop blocks to stabilize the loads, and manipulate switches, levers, and foot pedals to control the machines.

Hoist and winch operators work for logging companies, in construction, in manufacturing, and in mining and drilling for oil. For mining and drilling, they may set up derricks—tall, metal-cage towers that enclose a cable and pulley system. Hoist and winch operators also set up the elevators that many projects use to get workers and materials to the higher levels of the project (think about tall buildings or spaceships).

Although the machines do the heavy lifting and moving, operators need strength to use hand tools to move materials onto the equipment, reposition as needed, and use hand tools to tighten cables. Agility is helpful, as there may be a lot of climbing to make sure everything is safely secured. You should have a good sense of balance, distance judgment, eye-hand-foot coordination, and be comfortable with heights.

Another important part of the job is repairing and maintaining hoists, winches, tools, and other heavy-duty construction equipment. Hoist and winch operators have to be able to judge if a machine works properly, or if it needs a tune-up. They are also the troubleshooters when a machine stops working, finding the problem and figuring out a way to fix it.

Safety is always a concern for hoist and winch operators; necessary work safety gear includes hard hats and steel-toed boots, heavy-duty ear plugs or ear phones because they're always near loud machinery, and rain gear and sunscreen because the bulk of the job is performed outdoors. It's also critical that they know the hand signals to communicate quickly and effectively with coworkers who may be out of hearing range.

Some companies hire entry-level hoist and winch operators and provide on-the-job training. That hands-on training can last anywhere from one month to one year, depending on the operator, the machines, and the job. Once a new candidate has proven his or her ability to handle the machines and feels comfortable working independently, he or she may be asked to work on specific projects, or even train newer workers. Union apprentice programs are another way to learn the skills for these jobs.

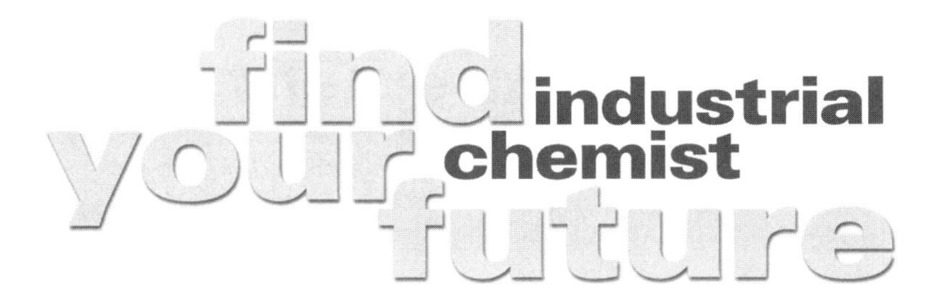

industrial chemist

Plastics. Cosmetics. Saccharin. These and other chemically-based products are a major part of our lives, thanks to the work of industrial chemists. These expert scientists use their knowledge of chemicals to turn raw materials into new products for consumers. Industrial chemists also work with products that consumers rarely (or never) see, but that are important parts of the manufacturing process itself that produces the products we do see.

Industrial chemists work on specific needs, rather than theoretical problems. Ink chemists, for example, work for large printing companies. They make sure that the ink used in printing will stay on the material (paper, cloth, or foil) that the printer is using and that colors are mixed correctly. They also develop special inks for special purposes. For example, a vegetable-based ink may be developed for a printed piece that will have to be licked and pasted on a form.

Industrial chemists study chemicals and how they work. They develop products and then run tests and experiments to make sure their creations are safe and cost-effective. Creating new materials is no easy task: industrial chemists need patience, attention to detail, and creativity.

Get Started Now!

- Load up on science and math courses, especially chemistry, physics, and algebra.
- Research manufacturing companies in your area that employ industrial chemists. Try to set up an informational interview with one of their chemists so you can get the low-down on day-to-day work.
- Participate in the American Chemical Society's annual Chemagination science essay and poster contest for high school students. For information click on the Educators and Students section of the ACS website at ***www.chemistry.org***.

Search It!
International Union of Pure and Applied Chemistry at ***www.iupac.org*** and the American Chemical Society at ***www.chemistry.org***

Read It!
Chemical and Engineering News at ***http://pubs.acs.org/cen/index.html*** and *ChemMatters* magazine via ***www.chemistry.org***

Learn It!
- B.S. degree in chemistry
- Master's degree preferred
- Ph.D. required for higher-level positions

Earn It!
Median annual salary for chemists is $52,900.
(Source: U.S. Department of Labor)

Find It!
Industrial chemists work for companies that manufacture products for pharmaceutical companies, the automotive industry, and other large sectors. See listings at ***http://chemindustry.chemjobs.net***.

Hire Yourself!

ChemMatters, the American Chemical Society's magazine for high school students, has invited you to submit a 500-word article for the next issue. The topic involves predicting a breakthrough related to chemistry that will improve the quality of a teenager's life. The prediction should involve one of four "hot" chemistry areas: biotechnology, health care, transportation, or the environment. Use the Internet to find background information to support your predictions.

Like other scientists, industrial chemists spend lots of time in a lab carrying out experiments and procedures. For this phase of the process, the chemists work with lab equipment—such as beakers and test tubes—as well as high-tech computers and high-powered microscopes. They keep careful, detailed notes about their experiments, making sure to record each step of the procedure and its outcome. Experimenting with a new material can take a long time: anywhere from several months to several years. Industrial chemists perform tests for two reasons. The first is to make sure their product is safe and effective before the material can be put to use. The second is to ensure that all safety provisions have been followed to ensure that using the product will not harm workers (no explosions when mixed with other chemicals) or end users (no harm will come to customers who buy things made with those chemicals).

During the testing phase, industrial chemists may come into contact with hazardous materials, liquids, and gases. To keep the environment and others safe, they may conduct experiments in an enclosed space. When doing this, industrial chemists may wear safety goggles, lab coats, rubber gloves, and other clothing designed for safety.

An important component of the industrial chemist's job is monitoring quality control in the plant. Chemists prepare detailed instructions for workers, telling them about the ingredients necessary for the products they are preparing, about proper temperatures, and about mixing or other prep times. Then they continually test samples of both the raw materials and the finished products to make sure that each batch is exactly the same, and that all product meets government standards.

Chemists are also involved in helping the manufacturing plant meet environmental requirements, ensuring that waste products do not exceed allowable levels.

Computer skills are important in this field, as is the ability to write clearly. Some understanding of marketing is a bonus, especially when it is time to promote the product to manufacturers. Although most companies have marketing departments specifically for that task, they may ask the chemist to attend meetings and speak with potential manufacturers.

Although you can get an entry-level position with a bachelor's degree, most industrial chemists continue their education to earn a Ph.D. As is generally the case, the higher the degree, the higher the salary. Industrial chemists with advanced degrees earn almost twice as much as their colleagues with four-year degrees. Be prepared to continually attend seminars and courses throughout your career: science and technology are constantly changing, and industrial chemists have to understand and apply those changes to their work.

Search It!
Industrial Designers Society of America at ***www.idsa.org***, Human Factors and Ergonomics Society at ***www.hfes.org***, and Chrysler Design Institute at ***www.chrysler.com/design***

Read It!
Design News magazine at ***www.manufacturing.net/dn*** and *The Professional Ergonomist* at ***www.bcpe.org/newsletters***

Learn It!
Minimum of a bachelor's degree in industrial design is typical.

Earn It!
Median annual salary is $52,260. (Source: U.S. Department of Labor)

Find It!
Industrial designers usually work for specialized design firms. Large companies often have industrial designers on staff. Almost one-third of all industrial designers own their own company.

industrial designer

Pick up a pen from your desk. Look at it carefully. Notice anything special about how it's designed? Most of us may think a pen is a pen is a pen. But for an industrial designer, the concept of a pen is a blank canvas waiting to be interpreted and reinterpreted. Pens can be retractable (or not), multicolor (or single), thick (or thin), with a smooth (or rough) texture. Pens can have any number of accessories attached to them—erasers, rubber grips, key chains, flashlights, or highlighters. There's even a ring-pen that slips over your finger to accommodate people with arthritis and helps prevent carpal tunnel syndrome (see it at www.ring-pen.com). And that's not even counting whether the pen is a conventional ballpoint, roller ball, felt tip, or gel. The number of decisions that go into the design of something as simple and as commonplace as a pen is staggering!

Industrial designers are responsible for making products attractive to customers, both in appearance and in utility. And every item that you see or use each day has been consciously designed by someone, from your computer (the screen, the keyboard, and the mouse), to your toothbrush (the handle, the bristles, the size, the shape), to the traffic lights in the street (the height, the diameter of each light, the exact hue and intensity of each color).

Get Started Now!

- Take classes in design and in technical drawing.
- Get curious about how and why all kinds of products are designed the way they are. Start noticing everything around you—from sports equipment to the kitchen sink!
- Part of figuring the right way to do anything involves learning to avoid the wrong way. Take a look at an on-line archive of bad designs at ***www.baddesigns.com*** to find out how not to design certain products.

Hire Yourself!

You've probably been sitting in them for years: those hard, uncomfortable chairs attached to a student desk. There must be a better way! Your job is to find it. Sketch out at least three ideas for providing comfortable, durable, and affordable classroom furniture. Keep in mind the ultimate design goal of creating classroom environments that are conducive to effective and enjoyable learning experiences.

Industrial designers also design things you don't see every day—medical equipment and instruments, manufacturing machines, construction tools, and even packaging for food, medicines, and cleaning supplies. Most industrial designers specialize in one area, whether it is consumer goods, home appliances, or medical supplies.

The industrial designer develops the design for a product, and prepares reports, drawings, and models to explain the idea to the client, including background research and manufacturing or shipping considerations. The client and the development team then review and refine the plan, until the designer can put the ideas into technical drawings and specifications to manufacture the item.

Some of the most important skills for industrial designers are creativity and curiosity. You must be interested in a wide range of topics and be able to see how regulations, trends, and changes in society influence the need for new products or changes to the existing products. You don't have to be an engineer yourself, but you should feel comfortable in working with the people who actually manufacture the products. Your wonderful new design will not be worth very much if it is exorbitantly expensive to produce or requires the purchase of totally new manufacturing equipment.

According to the Industrial Designers Society of America, industrial designers need the following five skills:

- Creative problem-solving skills
- Ability to convey concepts with quick sketches
- Good verbal and written communication skills
- Computer proficiency in vector-based or 3-D programs
- Mechanical aptitude and basic understanding of how things work

Working well as part of a team is also critical to being a successful industrial designer. In addition to coordinating with the manufacturing department, designers have to work with health and safety experts,

demographic researchers, model makers, market research specialists, and marketing and distribution teams.

Industrial designers find work related to a diverse range of specialties that includes transportation, medical products, consumer electronics, toys, furniture, environmental issues, entertainment, sports equipment, and many others. The opportunities seem limitless and, according to *Time* magazine, industrial design ranks as one of the 15 hottest professions in terms of job growth and compensation.

industrial engineer

"Industrial" usually refers to a manufacturing environment. Many industrial engineers do, in fact, work with heavy manufacturing processes. But industrial engineers also work for companies that provide services. They also work in the agricultural, health care, communications, and aerospace industries.

Engineers are usually thought of as people who do very detailed work with math, computers, and complex design questions. Industrial engineers may use some of those skills, but most of their work is with the "big picture," and with the people and processes that make the products or perform the services.

The primary goal of industrial engineers is to create a better product (or offer a better service) by using machines, materials, information, people, and energy more efficiently. They increase profitability by streamlining processes and eliminating wasted time or materials. Consider them the link between the present and the future—how things are done now and how things can be done more successfully.

This is a career for people who like to be on the go, as much of their time is spent in the actual work environments they are trying to improve.

Get Started Now!

- Take as many math, science, business, and technology courses as your high school offers.
- Hone your people skills! Get involved with clubs, organizations, or volunteer opportunities that put you in direct contact with various types of people and situations. Try to achieve positions of leadership; leading people and learning to influence their thinking will serve you well.
- Writing and communication skills are a must. Perfect these skills with extra writing classes, and become an active member of your school newspaper and debate team.

Search It!
Institute of Industrial Engineers at ***www.iienet.org***

Read It!
Industrial Engineer magazine at ***www.iienet.org/magazine***

Learn It!

- Bachelor's degree in industrial engineering
- Master's or more advanced degree required for some positions

Earn It!
Median annual salary is $62,150. (Source: U.S. Department of Labor)

Find It!
Industrial engineers work in virtually every sector of the economy, including manufacturing, telecommunications, agriculture, health care, and aerospace companies. They may also have positions within the government or educational institutes. Jobs are posted at the career center area of the Institute of Industrial Engineers website (***www.iienet.org/careercenter***).

Hire Yourself!

If your school is like most others, there's always room for a little improvement. With that idea in mind, your industrial engineering firm has been hired to improve the efficiency of some of the everyday functions of your school. What recommendations do you have for improving the efficiency of getting to lockers, how the cafeteria works, how classes are scheduled, or how the classrooms are arranged? Report your findings with a list of recommendations.

At the worksite, industrial engineers talk to everyone involved in the process, learn about each set of tasks and how they all interact with each other, and sometimes review the whole operation from a customer's point of view. On a typical workday, industrial engineers use people skills, business experience, and math and technology knowledge. Back in the office, industrial engineers may do research on how other companies or industries address similar situations to help them come up with solutions for their own company.

Industrial engineers must have excellent communication skills. They must be able to ask the right questions to get the needed information from workers. They also must be able to analyze their findings and present them clearly and persuasively to company management. They must be able to take a fresh look at things, be willing to get past "how it's always been done," and to work well both independently and as part of a team. They also need a strong background in technology to see how efficiencies can be implemented. Many industrial engineers tend to work more than a standard 40-hour week. How much more? That depends on where they're working. In manufacturing plants, they may be required to supervise extra shifts, even during the night. It's also not unusual for industrial engineers to carry their cell phone or beeper 24 hours a day. If a problem arises, they must be available to straighten things out. That's especially true for industrial engineers who work on a freelance basis—that is, offer their service to companies on a case-by-case basis. For freelancers, quite a bit of travel and extra work hours are the norm.

find your future industrial-organizational psychologist

industrial-organizational psychologist

Imagine a factory in which there is a room full of workers doing an extremely detailed, labor-intensive task. Imagine an efficiency expert advising the manager to increase the lighting in the room. What do you think would happen? Presto! The productivity of the workers increases. Imagine another efficiency expert coming in and recommending that they dim the lighting. If you think the productivity would decrease, you'd be wrong!

This experiment actually happened in a series of studies begun in 1924 at the Hawthorne Works of the Western Electric Company in Pennsylvania. The researchers found that whatever they did, the productivity of the workers increased. Their only hypothesis was that workers performed better when they felt that management cared about them and paid attention to them, a phenomenon that has since been called the Hawthorne Effect. Long after, other researchers found that when the novelty wore off, workers returned to their original levels, but the studies marked a change in how companies looked at their employees: as

Get Started Now!

- Take classes in psychology and statistics.
- Learn about basic business management. It will be helpful to understand the different ways in which companies are structured and the definitions of different management jobs that exist within most companies.
- Take a journalism class. As an industrial psychologist, you will have to write many reports to management. Journalism will help you learn to write clearly and efficiently.

Search It!
Society for Industrial and Organizational Psychology (SIOP) at ***www.siop.org***

Read It!
Industrial-Organizational Psychologist newsletter at ***www.siop.org/tip/TIP.html*** and *Journal of Applied Psychology* at ***www.apa.org/journals/apl.html***

Learn It!
- A bachelor's degree is required
- A master's degree is preferred
- A Ph.D. is recommended for those interested in research or teaching college

Earn It!
Median annual salary is $63,710. (Source: U.S. Department of Labor)

Find It!
For descriptions of currently available opportunities visit SIOP's JobNet at ***www.siop.org/JobNet***.

Hire Yourself!

Look on the Internet and in business and consumer magazines (such as *Fortune* and *Working Mother*) to find companies that have been designated as one of the "best companies to work for." Look for common denominators among these companies. What factors make for exceptional workplace satisfaction? Make a comprehensive list to use in helping an employer better understand the benefits of providing employee-friendly work environments.

individuals who would respond to their work environment, rather than just as living assembly lines.

The Society for Industrial and Organizational Psychology defines three goals of industrial-organizational psychologists: helping employers deal with employees fairly, helping make jobs more interesting and satisfying, and helping workers be more productive. The work mostly boils down to this: making the workplace a better place to work.

Some of the specific ways that these psychologists help companies are:

- Leadership and team building. Industrial-organizational psychologists use their knowledge of leadership and group dynamics to help companies organize their businesses, to help managers get the most from their departments, and to help people achieve the synergy of working as teams.
- Hiring and personnel management. Industrial-organizational psychologists help businesses find the best ways to identify the applicants who will contribute the most to the company. They help design interview questions and assessment tests, matching each person to the job that is right for him or her. Hiring well is critical at every level.
- Work environment. Industrial-organizational psychologists help businesses find the environment that will make their workers most productive. Whether it is the type of food offered in the cafeteria, the colors of the walls, the dress code, or different company benefits, the environment can have great impact on how well the company is perceived as a place to work, how long the employees stay there, and how productive they can be.
- Special groups. In every work situation, today's workforce is far more diverse than ever before. Industrial psychologists help companies become sensitive to issues raised by this diversity and to address these issues. For example: men and women may have

different responses to what behavior can be considered as sexual harassment. Industrial psychologists help create sexual harassment training programs that address the perceptions and behaviors of both men and women. Increasing ethnic diversity, the inclusion of people with disabilities, stereotypes of weight or appearance, and age discrimination are all areas that industrial psychologists work to better understand and to help companies address.

- Changes in the company. Industrial-organizational psychologists are often called upon to help companies deal with major changes. Massive layoffs or plant relocations must be presented in the way that best addresses workers' needs. Benefits packages must be appropriate to the displaced workers. Companies may even need to create programs to help the surviving workers keep their motivation and adjust to the new circumstances.

Organizational-industrial psychology is basically a technical field. Those who are not engaged in actively doing the research must still be able to keep up with new findings and understand the research that others have done. To do that well, it's important to have an understanding of research methodology and statistics.

Industrial-organizational psychologists often work for independent consulting companies or form their own companies. Some work for government agencies or large companies, particularly in the human resources area.

Search It!
American Arbitration Association at ***www.adr.org*** and Industrial Relations Research Association at ***www.irra.uiuc.edu***

Read It!
National Labor Relations Board at ***www.nlrb.gov*** and *Industrial and Labor Relations Review* at ***www.ilr.cornell.edu/ilrreview***

Learn It!
- A bachelor's degree is required
- A master's degree in business or labor relations and/or a degree in law are helpful

Earn It!
Median annual salary for labor relations specialists is $72,915. (Source: U.S. Department of Labor)

Find It!
Labor relations managers are found in large companies and in unions in almost every industry. See current job listings at the Monster.com website at ***http://labor.relations.representative.jobs.monster.com***.

labor relations manager

Are you the type of person that people come to when they need to settle a dispute? Can you see both sides of an issue, find a reasonable compromise, and get both parties to accept your suggestions? All of these qualities, plus excellent communication and listening skills, are critical for labor relations managers.

Labor relations managers (sometimes called industrial relations specialists) are the link between an employer's upper management and its employees (or the labor union representing its employees).

The field of labor relations is based on the existence of "collective bargaining" agreements—contracts or other agreements that are made between labor unions (or other employee organizations) and employers. In a collective bargaining process, professional negotiators work with management to develop rules, benefits, and procedures that apply equally to everyone in a job category. Employers save time and effort by being

Get Started Now!

- Take business math, business writing, or business law courses if they're offered in your high school. If not, pick up a book on the subject at your local library or bookstore.
- Participate in any peer mediation programs offered by your school or community.
- Hone your communication skills by joining the school debate club.
- Read about the National Labor Relations Board's (NLRB) co-op program for high school students, and their cooperative education program for college students at ***www.nlrb.gov/nlrb/about/careers/student.asp***.

Hire Yourself!

You're a labor relations manager at an automobile manufacturer whose employees are asking for a 10 percent salary raise and guaranteed overtime. Management of the company wants to give 2 percent raises, and cut overtime entirely. Present the compromise you would recommend along with the research (statistics on trends and economics) to back it up.

relieved of the need to negotiate with each employee individually, and employees benefit by having the force of the entire group behind them.

Contract negotiations are about much more than job responsibilities and salaries. They may include discussions of medical benefits, pensions, sick days and vacation days, and working conditions.

Labor relations managers work with both sides to negotiate a fair and equal deal for both management and employees. In order to accomplish this job, the labor relations manager does a lot of research on the assumptions behind the contract issues to see what makes sense. They use the Internet and other information sources to find statistics and trends that will support their recommendations.

For example, in negotiation of requests for salary increases, the labor relations manager may present statistics and research on changes in the general cost of living, average pay increase in the industry or comparable industries, and analysis of the profitability of the company or industry. That information can lead to compromises from either side. Management may see that their pay scale is lower than it should be. Or, employees may see that the company or industry is not profitable and agree to give up salary increases if the company promises not to lay off workers.

Since no one wins where there is a strike or other action that interrupts the functioning of the company, both sides in a contract dispute may agree to accept binding arbitration, the specific terms and conditions developed by a professional arbitrator (similar to an umpire or referee). Although each side loses a little negotiating power by agreeing to binding arbitration, both save the time and cost of court cases or strikes.

In addition to contract negotiations, labor relations specialists work with grievances (complaints filed by an employee, or group of employees, who feel they have been treated unfairly). Grievances can include issues such as unsafe working conditions, forced overtime, termination without required warnings and opportunities to improve, and unfair treatment (as when handicapped or women's bathroom facilities were not provided for those who needed them).

The most critical skills for labor relations specialists are excellence in communications, followed by diplomacy, patience, problem solving, and the ability to accommodate different viewpoints.

Many colleges or universities offer degrees in personnel, human resources, or labor relations. Your course work should include classes in science, statistics, social science, finance, law, management, labor economics or history, and, of course, a wide range of liberal arts classes. If you're hoping to advance to a management position, a master's degree is highly recommended. In fact, some companies prefer even their entry-level labor relations staff to hold a master's degree. In some positions, a law degree can also prove helpful.

manufacturing engineer

Manufacturing engineers apply their own special brand of problem-solving skills to take product ideas from prototype to mass-produced reality. The process begins with a manufacturing engineer and a model maker who collaborate to make a tangible example of the idea. This is called a product prototype. In some cases this process involves state-of-the-art technology called rapid prototyping. The most amazing of these rapid systems is called stereolithography—technology that engineers use to build 3-D plastic prototypes almost as easily as they could print something on a sheet of paper!

Once the prototype is designed and refined, the manufacturing engineer's job turns to figuring out the most efficient process for producing it. This is an especially challenging part of the job that can involve designing completely new manufacturing equipment or configuring existing equipment to do what it takes to do this particular job. The manufacturing engineer may also oversee the installation or reconfiguration of the equipment.

Get Started Now!

- Take classes in basic accounting and finance. Remember, one goal of the manufacturing engineer is to keep costs down. It will help to understand how costs are calculated.
- Learn about stereolithography at How Stuff Works at ***http://computer.howstuffworks.com/stereolith.htm.***
- Take one of the "virtual tours" of some manufacturing facilities at the Manufacturing is Cool site at ***www.manufacturingiscool.com/tours.htm***

Search It!
Society of Manufacturing Engineers at ***www.sme.org*** and Manufacturing is Cool at ***www.manufacturingiscool.com***

Read It!
Manufacturing Engineering magazine at ***http://tinyurl.com/29nz4*** and *SME News* at ***http://tinyurl.com/yudby***

Learn It!
- Bachelor's degree in manufacturing engineering
- Master's degree a plus

Earn It!
Median annual salary is $62,150. (Source: U.S. Department of Labor)

Find It!
Manufacturing engineers work in a wide variety of manufacturing environments. Jobs are listed at the Society of Manufacturing Engineers website at ***www.sme.org***.

Hire Yourself!

Look in your backpack and pick out one item. It might be a notebook, a calculator, a cell phone, or whatever. List all the steps involved in producing this item. Make a chart depicting a practical, efficient way for producing the item in mass quantity. Be sure to include a list of raw materials.

Sometimes this stage involves planning the most logical sequence for the production line. For some products, the steps don't matter. For example, you can produce the sleeves of a jacket, the front, and the back, in almost any sequence as long as they all come together at the end. The zipper, on the other hand, has to be at the end, when the pieces have all been made and stitched together and we know how the whole thing will fit together.

Manufacturing engineers consider many factors in designing the production line. Some components may need time to dry or "set" before they can be handled again, while others need to be used quickly, while they are still malleable. Some portions need to be in dust-free or temperature-controlled rooms. Manufacturing engineers may also be responsible for planning safety procedures and overseeing quality control.

Another important consideration involves determining the best use of resources: both human and mechanical. In both cases, the manufacturing engineer is responsible for making sure all resources are equipped to perform their assigned duties properly. With human resources, this may involve training, conducting skill-building exercises, or preparing very detailed written specifications that clearly convey each and every step of the process. With mechanical resources, a manufacturing engineer might use computer-aided engineering software and enlist the help of a computer-aided manufacturing (CAM) technician to program each piece of equipment to perform precise, perfectly-timed functions. When all resources—people and machines—are used efficiently, companies will experience as little downtime as possible.

Manufacturing engineers estimate production times and work with management to create realistic schedules. They also make realistic plans to accommodoate Murphy's infamous law that "if anything can go wrong, it will go wrong."

Manufacturing engineers don't stop with the production of the components and the final product. They also devise racks, bins, or other containers that can safely sort the components as they are being manufactured, and they help design the packaging that will be used for shipping the product to stores or warehouses and the packaging that will be used when the product is sold to the final consumer.

Many schools now have programs in manufacturing engineering. A typical program includes traditional and nontraditional (computerized) manufacturing processes, fundamentals of electronics and microprocessors, CAD (computer-aided design), CAM (computer-aided manufacturing), robotics, materials requirements planning, quality control, and engineering economics.

The watchword in today's manufacturing environment is "increased productivity," another way of saying that the name of this game is to produce more high-quality products in less time and with less cost. This, in a nutshell, sums up the essential role of a manufacturing engineer.

Search It!
Minerals, Metals, and Materials Society at ***www.tms.org*** and Materials Information Society at ***www.asm-intl.org***

Read It!
What is Materials Science? at ***www.strangematterexhibit.com/whatis.html*** and *What is MSE (Materials Science and Engineering)?* at ***www.crc4mse.org/what/Index.html***

Learn It!
- Bachelor's degree in metallurgical or materials engineering
- Master's degree in an area like metallurgy or ceramics engineering

Earn It!
Median annual salary is $62,590. (Source: U.S. Department of Labor)

Find It!
Find job postings at the following professional association websites: ***www.tms.org*** and ***www.asm-intl.org***.

materials engineer

Materials engineers work with all kinds of materials. And what are materials, you ask? Materials are the stuff that everything around us is made of. Take a quick look around you right now. Think about all the different kinds of materials in your locker, your closet, and your home. Put all the materials you find together and they still represent just a fraction of the more than 300,000 natural and man-made substances currently used in a mind-boggling number of ways.

According to the U.S. Department of Labor's *Occupational Outlook Handbook*, material engineers are involved in the extraction, development, and processing of the materials used to create a diversity of products. They work with metals, ceramics, plastics, semiconductors, and combinations of materials called composites to create new materials that meet certain mechanical, electrical, and chemical requirements.

Throughout history, materials have been so important that entire periods of civilization have been named for the materials that were most commonly used during that time—think Bronze Age and Iron Age. In the old days, people learned about the characteristics of different materials primarily through trial and error. They learned, for example, that they could increase the strength of metals by heating and cooling them. Today's material engineers use chemistry and physics

Get Started Now!

- Take courses in mathematics and chemistry. Most college programs will require these subject areas as prerequisites.
- Get involved with the Science Olympiad program. For information visit the Science Olympiad website at ***www.soinc.org***.
- Learn about some of the decisions that go into selecting the materials to make a bike at ***http://materials.npl.co.uk/NewIOP/TheBike.html.***

Hire Yourself!

Find out all you can about plastics at websites such as the Hands On Plastics website at ***www.handsonplastics.com*** and the American Plastics Council website at ***www.plastics.org***. Use your newfound knowledge to create a timeline describing the history of plastics. Be sure to include your own ideas for future uses of plastic in new products.

and sophisticated technology to learn about materials at the molecular and even atomic levels.

Materials engineers create new materials, develop new uses for known materials, and evaluate different materials to find the one most suited to meet design and performance specifications. The materials that manufacturers work with fall into five basic groups: metals, ceramics and glass, plastics and polymers, semi-conductors, and composites (which are combinations of materials, such as fiberglass, concrete, and particle board).

Although materials engineers may specialize in one of these areas, the processes they use are basically the same. Their first step is usually research. They use sophisticated computer models to determine the characteristics of the materials they are considering, and may also use X-ray diffraction and kinetic and thermodynamic theories.

At the design stage, materials engineers choose the best material for a particular product's purpose. By making the most appropriate choice, engineers can help products last longer, work more efficiently, accommodate human safety factors for both manufacturing workers and the end consumer, and protect the environment by cultivating earth-friendly practices. Certain house paints, for example, are purposefully designed to repel crayon marks and other stains. Various components of certain exterior paints make them ideal for use in various types of weather conditions. For instance, one type would be ideal for warm, damp, and salty coastal climates while another type would be better for places noted for extremely cold temperatures.

New materials have reinvented the equipment used in many sports, from baseball bats to snow skis and beyond. In the construction field, materials engineers find the right combinations of materials that won't crumble in an earthquake, won't melt in a fire, and will give tall buildings the flexibility to bend in the wind without cracking. In textiles, materials engineers have developed fabrics for competitive sports, space exploration, the military, and police and fire workers.

During the design stage, materials engineers use computer models to simulate and analyze the effects and potential problems of a product's design. They need to think through worst-case scenarios and ensure that the materials can handle the toughest of tasks. If they do find faults within the designs, they will work at creating solutions to prevent potential problems and, in some case, avert disasters.

Finally, materials engineers constantly test the products and materials they have designed or recommended. They usually work with a team of other engineers, technicians, and scientists and ensure that everything continues to run smoothly and safely.

Most materials engineers work in manufacturing industries, especially in computer and electronic products, transportation equipment, fabricated metal products, primary metal production, and machinery manufacturing. Materials engineers do a lot of their work on the computers in their offices. They also go to the facilities where materials are produced and to the factories or plants where their designs are being used. Future materials engineers can expect to find opportunities developing new materials needed for electronics, biotechnology, and plastics products.

find mechanical your drafter future

mechanical drafter

Mechanical drafters and designers combine their drawing skills with technical knowledge to create designs for all types of products. Their detailed sketches of furniture, automobiles, machines, tools, and other products provide the basic blueprints for manufacturing. These blueprints are very detailed and serve as working diagrams for the people who construct the items. Mechanical drafters include dimensions, fastening methods, and other engineering information in their work. Needless to say, mechanical drafters must enjoy drawing, as well as math and computers. They use these skills on a daily basis.

The expression "back to the drawing board" comes from the field of drafting. In the past, almost all drafting was done on a drawing board, and drafters often had to go back to make adjustments as product specifications changed. To manually prepare drawings, they used the same tools you may have used in geometry class: compasses, protractors, triangles,

Get Started Now!

- Enter technical and drawing contests to improve your skills. Visit the student resources section of the American Design Drafting Association website at ***www.adda.org*** for information about its annual competitions.
- Take courses in math, science, computer technology, design, or computer graphics, and any high school drafting courses available.
- Take as many art classes as your school offers.
- Practice persuasive speaking. You will have to present your ideas clearly and effectively to engineers, product managers, and customers.

Search It!
American Design Drafting Association at ***www.adda.org***

Read It!
American Design Drafting Association at ***www.adda.org*** and read the latest news about CAD technologies at ***www.cadalyst.com***

Learn It!
- Minimum of two-year degree in drafting
- Skill and experience in CADD (computer-aided design and drafting systems)

Earn It!
Average annual salary is $40,730. (Source: Department of Labor)

Find It!
Mechanical drafters and designers work for manufacturing companies, construction companies, and private engineering firms. Find out about job opportunities at the ADDA website at ***www.adda.org/jobbank.cfm***.

Hire Yourself!

It's back to the drawing board for you! You're working for an automobile company who wants you to create the next hot vehicle. Your task: design and draw a vehicle that takes advantage of natural resources (think: solar energy, electric power) without sacrificing speed, style, and technology. Create a detailed sketch of your car from all angles, making sure to include height, width, and weight specifications. Use colored pencils, watercolors, or other art supplies to color the different components of your street machine.

rulers, and other measuring tools. Today, drafters have the benefit of computer-aided drafting (CAD) and usually work on a very a high-tech drawing board: the computer screen.

That doesn't mean, though, that you should toss your sketchpad. Remember that CAD is still only a computer tool. Drafters must still have strong design skills and be able to do manual drafting when necessary. College drafting programs include courses in both manual drafting and computerized techniques. Plus, sketching a picture and drafting the specs require design work. And designing often stems from ideas sketched out on paper and colored with ink, airbrush, crayon, pencils, and other drawing supplies. In fact, the more kinds of media you're familiar with, the better your chances of landing a great job.

Mechanical drafters are also responsible for checking the dimensions of the materials to be used and assigning numbers to each of the materials (similar to the instructions that come with assemble-it-yourself furniture). The designs or blueprints that they create may be for an entire product (like a car) or for a part of a product (like components of the chassis). Their designs may be full-size, or drawn to scale (keeping the exact ratios of the parts the same as the full-size version).

Mechanical drafters have to be very familiar with how the final machine or system will work. They often have to provide schematic drawings or views from different angles to show how different components will relate to each other. They also lay out, draw, and reproduce illustrations for reference manuals and technical publications to describe the operation and maintenance of mechanical systems.

Drafters specialize in many different areas. Mechanical drafters may focus on die design, automotive design, tool design, mechanical equipment, or molds (in the plastics industry). Drafters in nonmechanical areas include architectural drafters, civil drafters, and electrical drafters.

mechanical engineer

Research, develop, design, manufacture, and *test* are verbs that describe the kind of work that mechanical engineers do. Whether it's a power-producing machine such as an electrical generator or internal combustion engine or a power-using machine such as a refrigerator or elevator, mechanical engineers create useful products that work for us.

Sounds pretty all-inclusive, doesn't it? In fact, this field is one of the broadest there is. Mechanical engineers design many of the products that we use in our daily lives—from automobiles and airplanes to kitchen appliances and prosthetic limbs. They also design the machines that are used in manufacturing those products, as well as the tools needed by other engineers. Some mechanical engineers work on a very large scale, designing, for example, the giant wheels and gears used in some factory machinery. Others work in the area of nanotechnology, creating high-performance materials and components by integrating atoms and molecules.

Get Started Now!

- Take advanced-level math and science courses. If your school offers AP classes and tests for college credits, make sure you take them.
- Take shop or mechanics courses to become familiar with blueprints, motors, tools, and mechanical equipment.
- Inquire about internship opportunities at a mechanical engineering firm in your community. You'll be able to help run the office while learning the trade.

Search It!
American Society of Mechanical Engineers at ***www.asme.org***

Read It!
Mechanical Engineering magazine at ***www.memagazine.org***

Learn It!
- Bachelor's degree in mechanical engineering
- Advanced degrees or prior work experience can be an advantage
- Licensing required in most states

Earn It!
Median annual salary is $62,880. (Source: U.S. Department of Labor)

Find It!
More than half of all mechanical engineers work in manufacturing. Others work for consulting companies or government agencies. Many mechanical engineering jobs are listed at ***www.asme.org/jobs***.

Hire Yourself!

Take a hint from the important mechanical engineers who came before you. Use the Internet or library resources to research some widely respected mechanical engineers. Use poster board to create a mechanical engineering "hall of fame." For each entry, describe the engineer and his or her major contribution to the field. To find information for some possible contenders, go to the American Society of Mechanical Engineers website at ***www.asme.org/education/precollege/achieve***.

A few questions can help you identify whether you have some of the most important natural aptitudes of a mechanical engineer:

- Are you mechanically inclined? People with inborn mechanical aptitude find it easy and interesting to take things apart when there's a problem and then put them back together again. Mechanical types are never intimidated by putting together a bicycle, a barbeque grill, or a do-it-yourself furniture kit.
- Are you really organized and detail oriented? "Close enough" is not acceptable in this field. Lives can be lost because of imprecise design or construction.
- Are you curious? Mechanical engineers have to think about everything that can possibly affect how their designs and products work in the real world—that includes things like temperature, altitude, which parts may wear out first, and even the weight of the product's user.
- Are you a good problem solver? Every project requires new solutions for new problems.

In addition, the ability to grasp complex math and physics concepts are critical to mechanical engineers. Computer skills are also a necessity. Both CAD (computer-aided design) and CAM (computer-aided manufacturing) are critical for designing products or processes and for conducting simulations that test a product's performance under many different conditions. Computers also help with scheduling, ordering materials, and calculating the costs of each project.

Good communication skills are another important trait. Even though mechanical engineers work with extremely complex information, in order to be effective they must be able to convey that information to other, less scientific people involved in the manufacturing process including clients, manufacturers, lawyers, and inspectors.

Many mechanical engineers become experts in specific industries such as automotive, power generation, or heating, refrigeration, and air-conditioning. Within each industry, individuals may specialize in one of several job areas.

- Research engineers start the process by looking at the goal and coming up with potential solutions. They may create a prototype, or experimental version, of the proposed solution.
- Design engineers take the theoretical ideas of the researchers and refine them into things that can be produced and marketed effectively.
- Testing engineers make sure products meet every requirement for quality and safety.
- Some mechanical engineers eventually gravitate toward maintenance, technical sales, teaching, or management.

Colleges that specialize in mechanical engineering provide ample opportunity for practice in this field. In addition to mechanical engineering theory, some schools offer—and encourage—work/study programs. These programs allow students to work with manufacturing companies to apply classroom work to real-life scenarios. That hands-on work experience gives students a big boost when it comes time to enter the workforce. In addition to an undergraduate degree and work experience, some states require students to pass an engineering ethics exam to become certified.

In the coming years, opportunities for mechanical engineers are expected to increase as the demand for improved machinery and machine tools grows and as industrial processes and machinery get more complex due to automation and robotics. According to the U.S. Department of Labor, emerging technologies in biotechnology, materials science, and nanotechnology will create new opportunities for mechanical engineers.

Search It!

Associated General Contractors of America at ***www.agc.org***

Read It!

Learn about the history of millwrights at ***http://mwron.tripod.com/history.html***

Learn It!

- High school diploma or equivalent, plus apprenticeship
- Engineering degree preferred by some companies

Earn It!

Median hourly earnings of millwrights are $20.19.
(Source: U.S. Department of Labor)

Find It!

Millwrights work for manufacturing companies, repair facilities, automotive companies, and labor unions, such as the International Union, United Automobile, Aerospace and Agricultural Implement Workers of America (UAW) at ***www.uaw.org***.

millwright

If you're the designated "fix-it" person in your family or the person the neighbors call on to assemble their gas grills, kids' bicycles, and "easy-to-assemble" furniture, you might find the career of millwright of particular interest.

Millwrights know the ins and outs of machines—and how to install them, assemble them, and repair them. They can see how the parts work together and make adjustments when things don't fit exactly right. In fact, the very job title comes from the days when these skilled workers did everything that was necessary to build mills—a mainstay of the American economy in earlier times.

Millwrights work on many things that are very familiar to all of us: cars, cranes, baggage claim carousels, roller coasters and other rides, monorails, escalators, and gaming machines. They also work on machines used in manufacturing processes: conveyors and material handling systems, robots, cranes, pumps, motors, fans, furnaces, turbines,

Get Started Now!

- Build up your toolbox. Make frequent visits to local tool stores to learn about tools and what they are used for.
- Take shop, mechanics, or machines courses in high school to learn how machines work.
- Take classes in mathematics, drafting, mechanical drawing, metal or industrial shop, and any courses that will help you become familiar with construction technology.
- Check your local library or bookstore for books on how things work (some good ones include Stephen Biesty's *Coolest Cross-Sections Ever* [New York: DK Publishing, 2001] and *The New Way Things Work* by David Macaulay (New York: Houghton Mifflin, 1998) to find detailed descriptions about how hundreds of machines and processes work.

Hire Yourself!

The local theme park is planning on putting in a new roller coaster. As millwright, you'll want to find out all you can about how roller coasters work and how they are put together. Get some basic information at the How Stuff Works website at ***http://entertainment.howstuffworks.com/roller-coaster.htm***. Then use your favorite Internet search engine (such as ***www.google.com*** or ***www.yahoo.com***) to seek out information about the world's best roller coasters. Using information you find on-line and your own imagination, create a step-by-step plan on how to go about installing a roller coaster.

dynamos, generators, compressors, concentrators, and coolers. They take care of the gears, pulleys, bearings, locks, valves, flywheels, limit switches, and dozens of other components.

Many millwrights appreciate the constantly changing landscape of where they work. Their skills may take them to an automotive factory one day and a water treatment plant the next. They work in hydro/nuclear plants, oil refineries, amusement parks, and food and beverage facilities.

Millwrights are on the scene from the very beginning of all kinds of building projects. When heavy machinery arrives at the job site, they unload it, inspect it, and move it into position. Because millwrights are well versed in the load-bearing capabilities of the various systems of ropes, cables, hoists, and cranes, they often decide what devices will be used and where. If their calculations indicate that the current flooring won't support the weight of the new machine, the millwrights figure out how to build a new foundation or reinforce the existing one. As you can imagine, there's quite a bit of math involved in figuring out the most appropriate method for each situation.

Millwrights also assemble new machinery based on the manufacturer's blueprints and instructions. They fit bearings, align gears and wheels, attach motors, and connect belts. They are also called in to maintain or repair older machinery, clean or replace various parts, and test the equipment once it is fixed. Because companies do not want to have extra downtime for routine maintenance, that work often occurs during nights, weekends, and holidays.

Millwrights are often needed to dismantle machinery, whether to gain access for maintenance and repair or to move it to another location. They need to have an excellent sense of organization—this is not the place for the person who takes an appliance apart, puts it back together, and finds that there are a few screws and bolts left over!

It's important for millwrights to have manual dexterity, reasonable physical strength, excellent hand-eye coordination, and the ability to manipulate small parts. Good eyesight is also critical.

Although the total number of millwright jobs is not expected to increase over the next few years, there is still excellent opportunity in this field. The average age in this field is higher than most, and many will be retiring in the near future. And, even though assembly of machines may continue to become more automated, maintenance and repair are still critical to companies that want to protect their investments.

nanotechnologist

What is the smallest measurement you can imagine? Cut this in half, and in half again, and again, and again. When you've done all that, you will still be a long way from the size of the materials that are used in nanotechnology!

Nanotechnology derives its name from "nanometre," a unit of measurement equal to one millionth of a millimetre. How small is that? Think of a human hair. Imagine dividing its diameter into 10,000 equal pieces. Each of those pieces would be one nanometer!

Like other professionals who blend scientific theory with the practical realities of marketplace product development, nanotechnologists are part scientist, part engineer, part researcher, and part inventor. Their work typically takes place in research and development laboratories found in a wide variety of corporations and government agencies. They rely heavily on computers and other technologies to innovate, develop, and test new product applications.

Get Started Now!

Let's get small. Here are some ways to start exploring the world of nanotechnology.

- Try to take advanced courses in physics, chemistry, biology, and computers. You need a broad base of knowledge to succeed in this career.
- Check out the Nanopicture of the Day at ***www.nanopicoftheday.org***.
- Get an overview of how nanotechnology works at ***http://science.howstuffworks.com/nanotechnology.htm***.
- Explore the Nanoworld at the University of Wisconsin site at ***www.mrsec.wisc.edu/edetc***. There's even a free book you can download called *Exploring the Nanoworld with LEGO*.

Search It!
NanoBusiness Alliance at ***www.nanobusiness.org***; Institute for Molecular Manufacturing at ***www.imm.org***, and NASA: Center for Nanotechnology at ***www.ipt.arc.nasa.go***

Read It!
Smalltimes at ***www.smalltimes.com*** and *Nanotechnology Now* at ***http://nanotech-now.com***

Learn It!
- Advanced degree in computer science, physics, chemistry, biology, or engineering
- For a list of colleges visit the National Nanotechnology Initiative website (***www.nano.gov/html/edu/eduunder.html***)

Earn It!
Median annual salary is $62,900. (Source: U.S. Bureau of Labor Statistics)

Find It!
Check out the nanotechnology business directory, NANOVIP, at ***www.nanovip.com/directory***.

Hire Yourself!

Go on-line to the ScienCentral News website at ***www.sciencentral.com***. Use the site's search engine to run a search on nanotechnology. Browse through fascinating articles you find there and use them as inspiration to come up with three ideas for ways nanotechnology might be used to improve the quality of life for teenagers everywhere. List your ideas on a poster or in a PowerPoint presentation.

The potential for this field is so great that at the end of 2003, President Bush signed the 21st Century Nanotechnology Research and Development Act, authorizing $3.7 billion for nanotechnology programs that will look for ways to detect and treat disease, monitor the environment, and produce and store energy.

Nanotechnology applications are already impacting lives. IBM researchers have been able to demonstrate a data storage system of one trillion bits per square inch—a device that would be able to store 25 million pages of data on a surface the size of a postage stamp. Materials behave differently at the nanolevel. Nanofibers are now used for some types of stain resistant clothing and nanoscale particles are used in many paints, catalysts, and other chemical products. L'Oreal, the cosmetics firm, uses this technology to transport nanosize ingredients such as Vitamin E to the skin.

Where is nanotechnology going? The Department of Defense is using nanotechnology to develop protective and adaptive fabrics for battlefield clothing. Imagine the advantages of lightweight, superstrong materials with special properties, such as incredibly sensitive sensors. According to *Small Times* (***www.smalltimes.com***), food companies, such as Kraft, are funding research to create food that can adjust its color, flavor, or nutrient content to accommodate each diner's taste or health condition. They see the potential for developing filters that can screen out specified molecules based on their shape rather than their size, making it possible to remove toxins or adjust flavors. They are even working on packaging that can sense when its contents are spoiling and alert the consumer.

Job growth predictions are especially rosy for nanotechnology. While there are approximately 20,000 people currently working in this field, the National Science Foundation (NSF) estimates that 2 million workers will be needed to support nanotechnology industries worldwide within 15 years. In this same time period nanotechnology-related industries are expected to generate over a trillion dollars of income.

Some scientists have compared pre-nanotechnology to trying to combine those Lego blocks while wearing boxing gloves. Pretty difficult! Nanotechnology lets us take our "gloves" off. Already NASA is making it possible for a molecular nanoassembler to create spare parts in space, and other scientists are investigating the generation of food by nonbiological means so that countries with limited resources can have adequate nutrition. In the retail segment, science is looking at ways that large retailers like Home Depot and Wal-Mart can use nanotube-enhanced displays that eliminate the need to physically change signs, relying instead on simple computer programming that will adjust displays throughout the store.

Nanotechnology can change the products we use and the way we manufacture almost everything. If you have great imagination, are excited by research, and capable of understanding advanced technologies, you may make a contribution to this rewarding and challenging field.

Search It!
Association of Professional Model Makers at ***www.modelmakers.org***

Read It!
"What is Rapid Prototyping?" at ***www.xpress3d.com/WhatIs.aspx*** and check out some of the uses for rapid prototyping (RP) at RapidPro Manufacturing at ***www.rapidpro.com/whatis.asp***

Learn It!
- High school diploma or equivalent, plus apprenticeship
- Postsecondary training opportunities listed at ***www.modelmakers.org/education_list.lasso***

Earn It!
Median hourly wage is $17.18. (Source: U.S. Department of Labor)

Find It!
Pattern and model makers work for machine shops, film studios, model maker firms, and manufacturing firms.

pattern and model maker

Think of model making, and most of us recall some sort of hobby kit to make model cars or airplanes. In the professional world, models can be small, like the cars in those hobby kits, or full size. Models are often used in the film industry. After all, it's much easier to film a small ship in a small pool with manmade waves than to create a full-size ship, go out into the ocean, and hope for a storm that will destroy it in just the right way.

In manufacturing environments, there are several uses for models. One is to ensure that a new product can work if it is made according to the plans of its designer. The model maker takes a two-dimensional drawing (or sometimes just a verbal description) and turns it into a version of the proposed product. It's really important that this model, or "prototype," be totally accurate and in perfect proportion to the final product. The model can help designers make sure that everything can really fit and work together. It can also be used to see how potential customers respond to the design.

To make a model, the pattern maker typically sets up and operates the lathes, drill presses, and other machines used to create the pattern, and then cuts, trims, and shapes it using hand tools and power tools. The

Get Started Now!

- Take classes in art and design, and math.
- Get some experience using hand and power tools by helping out around the house or working on a community project like Habitat for Humanity (***www.habitat.org***).
- Most pattern and model makers today work with CAD technology. Try to take a class that is an introduction to CAD.

Hire Yourself!

Go to the website of the patent grant and application section of the U.S. Patent and Trademark Office at ***www.uspto.gov/patft***. Click on "Quick Search" under "Issued Patents." For Field One, choose Title. For Term One, put in a common item—maybe bicycle, oven, or luggage. You'll get a list of all the submitted patents for improvements to the item you searched on. Find some interesting inventions. Use clay to create a 3-D model of that invention.

specific tools used are largely dependent on the material from which the pattern will be made. The pattern maker is also responsible for verifying that the pattern has exactly the same dimensions as were specified.

There are three basic techniques for model making:

- In an additive process, the model maker starts with a basic form of clay or wood, and adds detail to it. Think of starting with a lump of clay and adding on more clay for the nose, ears, mouth, and so on.
- In a subtractive process, the starting point is a hard material, such as stone, from which the unnecessary parts are removed by chiseling, cutting, grinding, or sanding. Think of Mount Rushmore—the sculptor had to remove every piece of the mountain that was not part of the faces he wanted to portray.
- Rapid prototyping is a process that sounds more like science fiction than reality. In the most basic of layman's terms, it is the process by which you can feed a 3-D blueprint of an object into an RP machine and automatically receive a model of that object made of metal, paper, or plastic.

In many ways the number one skill needed for success in this field is the ability to think outside the box. The profession requires both technical and creative expertise. Essentially, pattern and model makers are part designer, part inventor, and part skilled craftsperson. According to the Association of Professional Model Makers, the future of the profession depends on the synergy that connects technology with industrial design, architecture, and product development.

power plant operator

We all know how important electricity is in our lives, from heating and cooling our environments to cooking our meals and using our computers. If we had any doubts about our dependence on electricity, the huge blackout in August of 2003, which occurred from the Midwest all the way to the east coast and north into Canada, showed us just how dependent we are on electric power.

Electricity is generated at power plants, usually operated by utility companies using water (hydropower plants), oil or coal (fossil fuel plants), or radioactive materials (nuclear power plants). Power plant operators work in all these types of plants to ensure that energy is produced, turned into usable electricity, and delivered quickly and safely to homes, schools, workplaces, and communities.

A typical day on the job for a power plant operator includes "making the rounds" in the plant. During this walk-around, operators check out the boilers, turbines, generators, and auxiliary equipment through-

Search It!
Power Online at ***www.poweronline.com*** and North American Electric Reliability Council at ***www.nerc.com***

Read It!
Power magazine at ***www.platts.com/Magazines/POWER***

Learn It!

- High school diploma
- Associate's degree or college-level math and science courses
- On-the-job training or four- to five-year apprenticeship
- Certification from the NERC (North American Electric Reliability Council)

Earn It!
Median annual salary is $49,920. (Source: U.S. Department of Labor)

Find It!
Power plant operators work in power plants owned by utility companies or for independent power producers, manufacturing companies, and the military.

Get Started Now!

- Maintain high grades in your high school math and science courses, including algebra, chemistry, and physics.
- Call a power plant in your state and ask to set up an informational interview with one of their power plant operators. You may be invited on a behind-the-scenes plant tour.
- Beef up your computer proficiency! Operators and entry-level power plant candidates must know how to prepare reports, keep records, and track maintenance using computers.

Hire Yourself!

Learn more about how each type of power plant works. Check out your local library or bookstore, or the Internet. A good place to start is ***www.HowStuffWorks.com***. Look up hydropower plants, nuclear power, coal power, and power grids.

Make a chart comparing power generation by hydropower, fossil fuels, and nuclear energy. Compare the cost to build each type of plant, the cost of electricity, impact on the environment, and common risk factors.

out the plant to ensure that everything is working as it should be. If something is broken, leaking, dripping, or worse, operators either fix it themselves or bring in the appropriate maintenance workers. Power plant operators also distribute power demand among the different generators and combine the current from several generators to control the flow of electricity. They have to be alert to changing demands for power, starting or stopping generators and connecting or disconnecting them from circuits to meet the demands of the moment. They write reports on unusual incidents, keep logs of equipment maintenance, and use computers to record switching operations they initiate.

In modern power plants, with automated control systems, power plant operators may also specialize in specific functions. *Control room operators* work in a central control room. In older, non-centralized plants, *switchboard operators* control the flow of electricity from a central point, and *auxiliary equipment operators* operate and monitor valves, switches, and gauges throughout the plant. In a nuclear power plant, *reactor operators* control equipment that affects the power of the reactor.

The power plant operator's job can be compared to that of a fireman. Things can go along smoothly, quietly, and routinely for a long time, until something goes wrong. Then it's critical that each operator is logical, practical, and able to react calmly and efficiently in an emergency situation. They must all be team players, able to depend on each other, and communicate effectively. They need good problem-solving skills, good eye-hand coordination, and must be up-to-date on all applicable technology.

Power plant operators tend to stay in their jobs for a long time. The good news is that there is a lot of job security. The bad news is that there are relatively few new openings each year.

The best ways to get into the field are to come in as a formal apprentice or to work your way up from a related job. Although apprentice programs require only a high school diploma, two-year degrees in areas

like electrical engineering (or anything involving math, computers, and communications) are strongly preferred. Apprenticeship programs can last for four to five years and combine classroom education with on-the-job training. During this time, the apprentice is paid about 50 percent of regular wages.

In many cases, experience (with or without a formal apprenticeship) is valued even more highly than classroom education. You'll have a definite advantage if you understand the automated equipment, whether you start as a line crew worker, fixing electrical lines and other equipment, or as a power plant worker, firing boilers and controlling generators. To become a nuclear reactor operator, you'll also have to pass the Nuclear Regulatory Commission exams and take annual operation and written exams to maintain your license. In addition, all operators are encouraged to take refresher courses and go through simulation tests to keep up their skills.

printing press operator

When Johann Gutenberg conceived of the idea of printing presses in 1452, his immediate goal was to find a more efficient way to make copies of the Bible. Today, the market for books is quite a bit bigger . . . more than $26 billion dollars a year! And books are just a small part of the materials that come off of printing presses. There are newspapers, magazines, packaging materials (boxes, labels, bags, and shipping cartons), instruction booklets, telephone directories, business cards, postage stamps, paper money, and hundreds of other products.

Printing press operators are the people who actually work the machines that print text, designs, and illustrations. In addition to actually running the printing presses, operators also prepare the machines for printing, adjust them during a printing job, and maintain them for smooth operating. They make sure the color of the ink stays constant and that the

Get Started Now!

- Learn about the history of printing at ***www.communication.ucsd.edu/bjones/Books/printech.html*** or at ***www.techcolor.com/history.htm***.
- Set up an appointment for a tour of a printing press. While you're there, talk to an operator about the type of work he or she does.
- Volunteer to help out in your school office and ask to be trained on their copy machine. You may also want to check into volunteer opportunities at your favorite place of worship, for a political campaign of a candidate you admire, or with a nonprofit organization.

Search It!

International Paper: "Learn About Printing" at ***http://glossary.ippaper.com/ipaper/content/default.asp*** and "Manuscripts, Books, and Maps: The Printing Press and a Changing World" at ***www.communication.ucsd.edu/bjones/Books/printech.html***

Read It!

Printing News Online at ***www.printingnews.com*** and *American Printer* magazine at ***www.americanprinter.com***

Learn It!

- High school diploma
- On-the-job training and/or four-year apprenticeship

Earn It!

Median hourly wage is $13.95.
Source: U.S. Department of Labor

Find local printers through the Graphic Arts Information Network at ***www.gain.org*** and find out how printing works at ***http://entertainment.howstuffworks.com/offset-printing.htm***.

Hire Yourself!

The printing industry seems to have it own special language. Before you try to break into this field, learn to speak like a pro with a little help from an on-line glossary of printing and graphic terms found at ***www.printindustry.com/glossary.htm***. Create an illustrated dictionary of at least 15 to 25 new words. Include the definition and create or find an example that illustrates the word's meaning.

flow of ink to the rollers is just right. They continually check to ensure that paper is feeding smoothly, that margins and alignment are correct, and that there are no imperfections in the printed product.

Operators must be skilled in the specific process and type of press they run—and there are several types. Offset lithography, the most common process, transfers an inked impression from a rubber-covered cylinder to paper or other material. Most printing is done with offset lithography, and operators must be trained to work with either Web or sheet fed presses. Gravure printing is done by inking recessed parts on an etched plate, which is then pressed to paper to form the image. Flexography is a type of rotary printing, in which ink is applied to the surface by a flexible rubber printing plate with a raised image area. Electronic printing is one of the newer processes (also including electrostatic and ink-jet printing) that use a plateless or nonimpact process. Plateless processes are what we use when we go to quick printing or copy shops.

Even printing press operators, who work the machines independently, still have a lot of communication with others. In large printing environments, operators rely on other people who work at the press to supply the materials that need to be printed. Those who work with the largest presses also have assistants and helpers to work with them.

In smaller companies, operators may be responsible for a printing project from start to finish. That means everything from securing the materials from publishing houses or businesses, to collating printed papers as they come out of the press. It's a demanding job that requires keen attention to detail: every print job has to be of the highest quality.

Printing press operators need to have good mechanical ability so that they can adjust the press as needed and make required repairs. Some mathematical skill is required—operators have to compute weights and measures and be able to calculate the amount of ink and paper to do a job. As the printing industry becomes more technically advanced, the most in-demand press operators understand chemistry, electronics, physics, and color theory. They must also be able to communicate clearly, both verbally and in written form.

Printing press operators need to have at least a high school diploma. And whether they go on to learn skills through on-the-job training, industry-sponsored apprenticeships, or college-level courses, operators who can think on their feet and react quickly will be highly valued, due to all of the advances in automation in the printing industry.

Search It!
Educational Society for Resource Management (formerly the American Production and Inventory Control Society) at ***www.apics.org*** and American Management Association at ***www.amanet.org***

Read It!
APICS magazine at ***www.apics.org/Resources/Magazine***

Learn It!
- Bachelor's degree in business
- Advanced degree in engineering management

Earn It!
Average annual salary for industrial production managers is $67,320. (Source: U.S. Department of Labor)

Find It!
Production managers work in commercial printing plants, metal fabrication shops, and manufacturing companies that produce automobiles, electronic components, and plastic products. Find examples of job listings at the APICS website at ***www.apics.org/resources/careercenter/jobseekers.htm***.

production manager

Every day you are likely to use dozens of manufactured products—from the quick snack you grab on the way out the door in the morning, to the chair you sit in at school all day, to the wheels you ride on to get home every night. Many of these goods—and millions of other ones—are produced in manufacturing plants within the United States under the supervision of an industrial production manager. Industrial production managers play a key role in making sure the products are produced in a timely and efficient manner. In other words, they're the ones to thank when the products you want are available at your favorite stores and when the quality of the products is worth the money you spend on them.

Industrial production managers work in all types of manufacturing plants, including industrial machinery and equipment, transportation, electronics, metal products, and food and beauty products. No matter what sector they work in, production managers have similar duties. An

Get Started Now!

- Business administration and accounting classes will give you a strong foundation for this field. Get involved with your school's business club, or start one up if your school doesn't offer one.
- Inquire about internships that can help you improve your business, management, scheduling, and planning skills. Companies that track inventory—such as car dealerships or grocery stores—can offer great experience.
- Join the debate club to hone your negotiation and persuasion skills. Managers use these on a daily basis.

Hire Yourself!

According to APICS Education and Research (E&R) Foundation, the following topics describe the types of issues that matter to industrial production managers:

- Inventory management
- Just-in-Time production
- Logistics
- Manufacturing processes
- Master planning
- Material and capacity requirements planning
- Systems and technologies
- Supply chain management
- Total Quality Management

Choose one and use a favorite Internet search engine (such as ***www.google.com*** or ***www.yahoo.com***) to find out more about it. Use the information you find to write a position statement about how your understanding of the issue might help you in your first job as an industrial production manager.

average day on the job might include planning a production schedule, hiring people to meet that schedule, selecting machines to do the job, analyzing and testing product quality, and keeping track of inventory. Plus, they typically are in charge of coordinating production activities with those of other departments. It's a job with varied responsibilities and one that brings new challenges each day.

A production manager's job pretty much boils down to two things: time and money. Their ultimate goal is to make the most of both.

Thus, production managers don't have time to chat by the office water cooler all day. They're too busy making sure the plant and all the departments are working optimally to meet the production schedule. That means production managers are constantly communicating with plant workers, the purchasing department, suppliers, and inventory workers to ensure all areas are running smoothly and on schedule. At the end of the day, production managers may have to write and review reports and analyze production data for the coming days. They make necessary adjustments to schedules, deadlines, budgets, and staffing to help them meet production goals.

Since the nature of the work is time-sensitive, it's not unusual for production managers to work more than a standard 40-hour workweek. When a schedule deadline looms, managers may even work round-the-clock. And in the event of a production emergency, managers may be called in to resolve the problem—no matter what time it happens and how long it takes to fix. In addition, managers have to deal with worker-related problems that may take lots of time and energy to be resolved.

To start out in this career, most manufacturing companies require production managers to have at least a four-year college degree with an emphasis in business administration, industrial technology, or industrial engineering. Some companies prefer managers to have a master's degree in industrial management or business administration. Excellent problem-solving skills and interpersonal skills are top priority, as are communication skills. Managers new to a company may also attend the company's training program to become familiar with company policies, schedules, and other requirements. The education doesn't stop there: production managers also take continuing education courses and attend professional seminars and trade shows to learn about technological and work-related changes in their field.

Industrial production managers held about 182,000 jobs in 2002, according to the Bureau of Labor Statistics. That number is expected to grow slower than average through 2012 because of increased productivity due to computers and planning. In addition, budget cuts have put more responsibilities on production managers—and that, in turn, has cut down the need to create more jobs. However, a number of job openings will stem from the need to replace workers who transfer to other occupations or leave the labor force.

purchasing agent Are you a born shopper? Do you love snagging great deals? Can you spot a trend months before it hits the stores? If you answered yes to any or all of these questions, you may find your career calling as a purchasing agent. Purchasing agents buy goods and services for companies. Their mission is to buy the highest quality merchandise at the lowest possible prices. Interested? Read on.

As a purchasing agent, you'll determine what types of items to buy from suppliers and try to negotiate the lowest price. Think of it as haggling for a product—but in a much more sophisticated way. In fact, purchasing agents often work out detailed international contracts for the products they buy. The contracts must outline the exact items to be purchased, the cost of the items, any materials that need to be bought, and any equipment that's required. Drawing up contracts is tricky; purchasing

Get Started Now!

- Find out which retail and wholesale stores in your town or city use purchasing agents. Talk to supermarkets, drugstores, building companies, and manufacturing companies.
- Take marketing, business, or communications courses offered in your high school.
- Keep track of trends. Stock up and read through magazines specific to the field you're interested in. If it's fashion, pick up *InStyle* or *Women's Wear Daily*. If it's automotive, read *Car & Driver* or a weekly car publication. Find journals, magazines, and newsletters that give you a heads-up on trends.

Search It!
Institute for Supply Management at ***www.ism.ws***

Read It!
Purchasing Magazine Online at ***www.purchasing.com***

Learn It!
Two-year degree in marketing management or four-year marketing degree with an exam and certification in purchasing management are typical.

Earn It!
Median annual salary is $45,090. (Source: U.S. Department of Labor)

Find It!
Purchasing agents work for companies that want to purchase goods, including large retail stores, such as Macy's (***www.macys.com***), and communications companies, such as AT&T (***www.att.com***). Purchasing agents also work for local, county, state, and national government departments.

Hire Yourself!

Be a purchasing agent right now—for your home. Raid your home pantry, closet, and desk to make a list of 15 products your family uses on a regular basis. Then go online, look in the Sunday newspaper, and check other resources to find three places that sell the same items. Compare prices and other special offers. Use a highlighter to indicate the best source for purchasing each item.

agents walk a fine line between satisfying the supplier while meeting the needs of the company they work for.

Other aspects of the job include working with marketers to boost product appeal and keeping up with supply and demand. If a product sells quickly—and unexpectedly—purchasing agents must be ready to get that product back to their company. And fast. Say, for example, Big Retail Company A buys shoes from Supplier B. In the meantime those shoes have gotten wildly popular so when they hit the store, they fly off the racks. If Retail Company A can't order more of those shoes from Supplier B, they've got a huge financial and customer-service problem on their hands. A big part of a purchasing agent's job is making sure those problems don't occur. That means hammering out contracts that cover all bases and finding suppliers who can deliver on demand.

All companies use purchasing agents—from furniture stores to cell phone companies to retail clothing stores, and more. Purchasing agents—also known as buyers and merchandising managers in some industries—who work for wholesale or retail firms usually buy finished products to sell to consumers. That means purchasing agents have to know what consumers will want in the near future. How do they get the inside scoop? Purchasing agents spend lots of time reading trade magazines, attending trade shows, and watching media. They also track consumer spending and buying patterns. Combined, this information gives purchasing agents a good idea of what consumers will desire or need.

If you love clinching a deal and don't mind working hard to do it, you'll do well in this field. Purchasing agents often work more than 40 hours per week, especially when sales, production schedules, and contracts need extra attention. A fair amount of time is also spent on the road, traveling between supply companies, trade shows, and the company they work for. And because they're constantly meeting, talking to, and negotiating with others, purchasing agents need excellent communication skills.

Would-be purchasing agents have to work their way up the ranks. Companies usually hire graduates as purchasing clerks. From there, they may work their way up to assistant buyer and other mid-level jobs. Those years of training allow employees to become familiar with the terrain and help them build strong contacts in the field. Several years of good experience are required before getting the management role of purchasing agent.

If you're interested in becoming a purchasing agent, there are several routes you can take. Two-year programs in marketing management coupled with work experience provide excellent training for work in many retail and wholesale establishments. For students with a degree or interest in marketing, a four-year marketing degree can be combined with purchasing manager exam certification. Still another route is to study procurement and materials management at a four-year college or university. Not sure what you should do? Choose a few companies you'd like to work for and find out what education they require for purchasing agent positions. This simple exercise will give you a starting point in this exciting career.

Search It!
American Society for Quality at ***www.asq.org***

Read It!
Quality Progress at ***www.asq.org/pub/qualityprogress***

Learn It!
- Two-year technical degree for technicians
- Bachelor's or master's degree in engineering for engineers
- Certification is desirable for both technicians and engineers

Earn It!
Median hourly wage is $13.01. (Source: U.S. Department of Labor)

Find It!
Quality control technicians usually work for manufacturing companies, though most customer-driven businesses have quality assurance engineers. Major employers include Boeing at ***www.boeing.com*** and General Motors at ***www.gm.com***.

quality control technician

If you're a perfectionist who insists on the highest quality, a career as a quality control technician or quality assurance engineer will put that characteristic to good use. They are the ones responsible for making sure the products you love live up to your high standards.

So what's the difference between a quality control technician and a quality assurance engineer? Schooling, for one. Engineers must have a bachelor's or master's degree in engineering and pass quality control exams. Technicians, on the other hand, are only required to take two years' worth of technical courses and obtain their associate's degree. On the job, engineers often conduct scientific and mathematical experiments. Technicians take a more hands-on approach with tools and visual inspections.

No matter what their schooling, people who work in quality control have to make sure the products they're responsible for meet or exceed quality standards. How do they do it? That depends on where they work. Job duties vary from company to company and across different industries.

Get Started Now!

- Use the Internet to find a quality assurance association or professional organization in your state. Ask if you can attend a meeting, and use that opportunity to speak with the members about their career path and job responsibilities.
- Become familiar with terms and tests through the International Standards Organization at ***www.iso.ch***. It issues quality control standards to companies.
- Take computer classes through your high school or through community programs, such as the YMCA.

Hire Yourself!

You've been called to interview for a position as quality control technician for a company that manufactures shoes. To get acquainted with the types of quality issues associated with shoes, head for your closet. Pick out three pairs of shoes. Analyze each shoe's quality according to the following criteria: comfort, appearance, durability, quality of materials, and workmanship. Record your findings on a sheet of paper divided into three sections—one for each pair of shoes.

Quality control technicians and engineers often make detailed visual inspections of the product being manufactured. If a company manufactures chocolate, quality control technicians look at, smell, and—yum!—taste the confections to make sure they're rightfully tasty. A quality control engineer working at an automotive company, for example, may make visual inspections to look for scratches, color changes, and part defects. Yet a quality control technician working for another automotive company may check equipment and instruments. Again, job responsibilities vary widely.

So, what do you need to prepare yourself for this line of work? Start with beefing up your computer skills. Many companies have switched from humans to computers and automated machines to check for quality standards. That means job opportunities have declined—and will continue to do so, according to the *Occupational Outlook Handbook*. Your best line of defense: learn how to run, control, test, and fix those machines. Techs and engineers with that type of background will often get first dibs on a job.

Another crucial requirement for this field: attention to detail. Quality assurance engineers and technicians must zone in on the smallest details. That means good eyesight is an absolute must—with or without glasses or contacts. In addition, people working in this field require manual dexterity and good coordination to use tools necessary to get the job done.

If this career peaks your interest, you may also want to look into a career as an inspector, tester, grader, sorter, or sampler. These people do exactly what their titles imply. They compare color, shape, texture, size, weight, and more in order to do their part in quality control. In large companies, they're part of a quality control team that's often headed up by a quality control engineer.

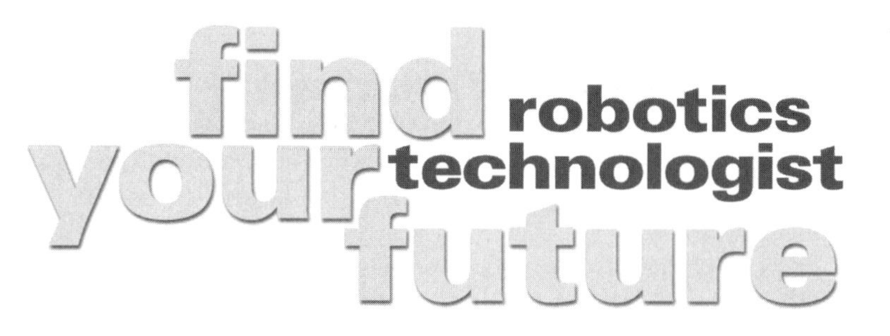

Search It!
Robotics Industries Association at ***www.robotics.org***

Read It!
Machine Design at ***www.machinedesign.com***
Robotics Online electronic newsletter at ***www.robotics.org***

Learn It!
- Technology diploma from a technical institute
- Advanced degree in engineering with a specialization in robotics technology

Earn It!
Average annual salary is $40,020. (Source: U.S. Department of Labor)

Find It!
Major companies include Kawasaki Robotics (***www.kawasakirobotics.com***), Panasonic Factory Automation Company (***www.panasonicfa.com***), and Epson Robotics (***www.robots.epson.com***).

robotics technologist

Most manufacturing companies have turned to automation—computerized machines and equipment—to get various and repetitive jobs done. Those robotic and computerized machines have taken jobs away from countless people—except robotics technologists. Robotics technologists create high-tech robots and equipment used in many homes, businesses, and industries.

Robotics technologists are responsible for designing automated equipment and robots and creating the programs to run them. Once the robots are in production, technologists supervise the progress and keep track of their assembly. Technologists are also responsible for updating and redesigning parts to keep up with the fast pace of technology.

Get Started Now!

- Sign up for and become an active member of your school's computer or electronics club. If your high school doesn't offer one, seek out community organizations that focus on computers, technology, or robots.
- Build your own robot! Check a computer store for software designed for this purpose. It will give you a good foundation in robotics technology. Visit the Robot Directory website for inspiration at ***www.robotdirectory.org***.
- Look for internship or apprenticeship opportunities through local companies that specialize in robotics, electronics, or mechanics. This type of hands-on experience will prepare you for the field.
- Keep up with robotics industry news through websites and robotics magazines.

Hire Yourself!

You're a robotics technologist who has just been chosen for a secret mission: create a tool that can be used to help the federal government. Your robot can be used in any of these areas: space, environment, espionage, or communications. Design your robot on a piece of paper. What components does the robot have? What does it do? How does it work? Who can use it? Next, search the Internet to find out what types of robots are used in the field you chose. How is your robot design similar to or different from robots that already exist?

Some robotics technologists work in a research and development laboratory on the design, programming, and production end. Others prefer to maintain, repair, and upgrade machines that are already in use. And some technologists use their skills to sell the robotic equipment to companies. This field offers a wide range of options—it's just a matter of choosing the one you prefer.

Where are robots used? Virtually everywhere. You may think robots are high-tech machines created to mimic human tasks and behavior. Although some robots that emulate humans do exist, the majority of robots are electronic machines designed for specific jobs. The Mars rovers *Spirit* and *Opportunity*, for example, were robots designed to head to the red planet and take photos of its environment. JASON was a robot used to probe the ocean floor. Those are just two examples of what robots can do: go to heights and depths humans can't easily reach. These types of robots help humans gain a greater understanding of the world.

Still other robots are delegated to the mundane—yet equally important—tasks of assembly-line manufacturing. Robots like these may load, sort, cut, and press components used to manufacture a product. Large manufacturing companies rely on these types of robots to keep costs down and performance up. Whatever their role, robots have become an increasingly important part of life. Without robot technologists, these sophisticated machines would not exist.

What does it take to create, design, and run robotic equipment? A love of computers, for one. Most technologists spend at least half of their day in front of a computer screen, especially if they've chosen to work in design and programming. Math and science skills are also a plus. Technologists count on mathematics and scientific principles to do their job. In addition, knowledge of mechanics and electronics is necessary. Sound daunting? Don't worry. You don't have to be perfectly skilled in each of these areas. General courses and training provide technologists with enough knowledge to work effectively in this field. A specialization in the specific area you'd like to work in—such as mechanics or computers—will prepare you for a job in the field.

A diploma from a technical institute is a great way to get started in this field. For those seeking further education, a bachelor's degree in engineering with a specialization in robotics will help prepare you for the field. Master's or doctoral degrees in robotics technology are a good idea for expertise and advancement in the field.

semiconductor processor

Since the first microchip was invented in 1974 to be used in hearing devices, the semiconductor industry has exploded. According to the Semiconductor Industry Association (SIA), the industry has manufactured about 90 million chips for every person on the planet. By 2010, that number will grow to 1 billion per person. In addition, the U.S. Bureau of Labor Statistics ranked the semiconductor industry second in the nation for the lowest injury and illness rate out of 208 industries. These kinds of statistics make semiconductor processors really happy.

If job openings, job security, and job safety are tops on your list, a career as a semiconductor processor may be exactly what you're looking for. These trained and skilled people make semiconductors, which are tiny circuits on a silicon wafer. One of those wafers contains hundreds of circuits called microchips. These microchips are used nearly

Get Started Now!

- Visit a semiconductor plant and talk with the technicians who work there. Ask them about their workload, schedule, and working environment. You might even ask for a tour of the clean room.
- Sign up for a newsletter related to semiconductors, and keep track of industry trends and statistics. A good place to start is the Internet or statewide semiconductor organizations.
- Find out what types of products use semiconductors and how they're made. Search Semiconductor Online at ***www.semiconductoronline.com***.

Search It!
Semiconductor Industry Association at ***www.sia-online.org/home.cfm***

Read It!
Semiconductor Online at ***www.semiconductoronline.com***

Learn It!
Two-year technology degree in semiconductor technologies is typical.

Earn It!
Median hourly wage is $13.14. (Source: U.S. Department of Labor)

Find It!
Semiconductor processors work primarily for semiconductor fabrication facilities. Large employers include the National Semiconductor Company (***www.national.com***), Advanced Micro Devices (***www.amd.com***), Micron (***www.micron.com***), and Integrated Device Technologies (***www.idt.com***).

Hire Yourself!

According to the Bureau of Labor Statistics, the electronic component and accessories industry will be one of the most rapidly growing manufacturing industries in the next decade. To keep the industry booming, manufacturing companies are making semiconductors smaller, more powerful, and more durable. How do they do it? Investigate the tech changes in semiconductors and microchips. What new materials will be used? How will manufacturing become more streamlined? How advanced will future semiconductors become? Next, imagine how those super-fast semiconductors will impact daily life and businesses. Write a one-page briefing paper describing the changes and their implications.

everywhere: computers, cell phones, TVs, microwaves, machines, traffic lights, and much more. You probably have never seen one up close, but your daily schedule depends on these microchips to work.

What's the recipe for a semiconductor? Processors use crystal growing materials and specialized equipment and processes. First, they use photolithography, a printing process that creates plates from photographic images. Next, they use high-tech equipment to etch microscopic patterns of the circuits onto wafers. They then press metals that conduct electricity into the patterns. Next, they apply chemicals to the wafer to make the circuits smooth. Finally, they repeat the process again on a new layer—up to 20 times! Each semiconductor has from eight to 20 layers of circuits.

All of this takes place in semiconductor fabricating plants, known in the business as "fabs." Inside those fabs, manufacturing takes place in "clean rooms"—which are exactly what they sound like. Clean rooms must be kept free of dust, because dust can damage semiconductors. This is why processors must wear "bunny suits" over their clothing to keep lint and dirt away from the chips. For the same reason, access in and out of the clean rooms is controlled and limited. Each time a worker exits the room, he must be suited up in a new, clean suit and decontaminated before entering again.

There are lots of people involved in creating semiconductors. Operators power up and monitor the equipment used in the production process. Technicians are responsible for quality control and maintaining equipment. They keep an eye out for flaws or damage in chips and track down where it came from. If damage was caused by equipment, they must run diagnostic tests to figure out what caused it and how to fix it. Although technicians and operators may work independently at times, they usually work in teams to keep production flowing smoothly. Those teams work eight- to 12-hour shifts, depending on the plant. And because semiconductor plants operate 24 hours a day, night-shift work is standard for many workers.

If you're interested in a career in the semiconductor industry, beef up your math and science skills. They're essential for this type of work. Communication skills are key, too, because operators and technicians must be able to convey information to each other in a clear and timely manner.

If you live near a semiconductor plant, you may be able to apply for a summer or part-time job. Although most companies require employees to have a two-year technical degree, they may offer on-the-job training for students who want to get started in the field.

sheet metal worker

Across the United States, hundreds of industrial, commercial, and residential buildings are constructed each day. It's no wonder, then, that sheet metal workers are—and will continue to be—in very high demand. These skilled laborers do everything from producing, installing, and maintaining heating and cooling systems to creating restaurant equipment to making outdoor signs. Roughly two-thirds of sheet metal workers are employed in the construction industry. But one out of three workers is employed in the manufacturing sector, where the key concern is mass producing sheet metal products.

In manufacturing plants, sheet metal workers make sheet metal parts for large-scale products, such as heavy-duty equipment and aircraft. They work with automated industrial equipment and may be required to maintain those machines. Mostly, though, manufacturing sheet metal workers use fabricating machines to cut, bend, and straighten metal. They also use other tools such as anvils, hammers, and soldering and welding equipment. Needless to say, this job is physically demanding—and comes with a dose of job-related dangers.

Get Started Now!

- Take algebra, trigonometry, and geometry classes—they'll come in handy for work with a small shop where workers draft, design, and cut sheet metal.
- Some states require certification before you can do work. Check with your state's labor department for sheet metal regulations.
- Get in on the ground floor! Take advantage of local sheet metal shops in your area. Ask about summer jobs, part-time shifts, or apprentice opportunities. Sheet metal workers who own their own shops may be willing to train you.

Search It!
Sheet Metal and Air Conditioning Contractors' National Association at ***www.smacna.org*** and Sheet Metal Workers International Association at ***www.smwia.org***

Read It!
Architectural Metal at ***www.naamm.org***

Learn It!
- Four- to five-year apprenticeship, plus 144 hours per year of classroom training
- Information about local training opportunities available at ***www.sheetmetal-iti.org/contactsearch/NorthAmericamap.asp***

Earn It!
Median hourly wage is $16.62. (Source: U.S. Department of Labor)

Find It!
Sheet metal workers are employed by companies like Apollo Sheet Metal (***www.apollosm.com***) and NewMech Companies, Inc. (***www.newmech.com***).

Hire Yourself!

Do you have a favorite kind of roller coaster? You can thank sheet metal workers for creating them! How do they do it? Search the Internet to locate information about your favorite roller coaster. (A good place to look is any website or group devoted to roller coasters!) Next, look for stories or statistics about those coasters and how they were made. Investigate further: What kind of metal was used, where was it produced, how long did it take, and how many workers were required from coaster conception to finished speed machine? Make a timeline or photo essay (use photos from books or the Internet) to tell the story of your favorite roller coaster—and how sheet metal workers were key in its development.

If you've ever seen a cut or torn piece of metal, you know it can be razor-sharp. Sheet metal workers have to be cautious when working with metal so they don't get a nasty cut. When welding or soldering, they also have to be wary of serious burns. And because the equipment they work with operates at high speeds, workers must always be mentally and physically alert. Whether they work in a factory, small shop, or in commercial or residential buildings, sheet metal workers take care to follow safety procedures.

Sheet metal workers usually suit up for the job with safety goggles, steel-toed boots, breathing masks, and hard hats. They also check the mirror before heading to work: jewelry or loose-fitting clothing are definitely not allowed because they can get caught—and trapped—in a machine.

Learning this trade takes dedication, time, and a hefty helping of mechanical and mathematical skills. Some large-scale manufacturing shops allow people to start off as helpers to experienced sheet metal workers. That means they may at first only clean up debris, but later learn to operate tools and machines. Although many shops prefer a high school diploma or equivalent, smaller shops may hire workers with great mechanical aptitude and a strong desire to learn the trade.

Another route for this career is a four- or five-year apprenticeship program offered through the Sheet Metal Workers' International Association and local chapters of the Sheet Metal and Air Conditioning Contractors' National Association. The apprenticeship hooks you up with local sheet metal shops, installers, or production factories. On the job, apprentices learn how to work with sheet metal, create pattern layouts, and make metal components for air conditioning and heating systems. Classroom

training consists of drafting, mathematics, welding, computerized equipment, and the basics of heating, cooling, and ventilation systems. In all, apprentices attend approximately 144 hours of classroom training each year—and training doesn't stop there.

Experienced sheet metal workers keep up with technological changes in the field, as well as new types of advanced machines. Many workers attend training courses offered through a union or their employer. The additional knowledge may give employees a boost when applying for supervisor positions or specialized training.

If you're seriously considering this career, now is a good time to get to work! The Bureau of Labor Statistics expects job openings to increase through 2012, reflecting growth in the demand for sheet metal installations as more industrial, commercial, and residential structures are built. But don't wait until then to start training. High school courses in algebra, geometry, trigonometry, and machine shop will get you prepped for this career.

tool and die maker

Even though it's unlikely that you know what they make, it's quite likely that, no matter where you are, you are surrounded by the handiwork of tool and die makers. According to the U.S. Department of Labor's *Occupational Outlook Handbook*, these professionals are among the most highly skilled workers in manufacturing. Their job is to produce tools, dies, and special guiding and holding devices that enable machines to manufacture a variety of products used in everyday life—from clothing and furniture to heavy equipment and parts for aircraft. Their job requires a high level of precision, concentration, and skill.

Tool and die makers are often considered highly specialized machinists. And, just for clarity's sake, you should know that there is a distinction between toolmaking and die making, although many tool and die makers know how to do both. Toolmakers produce the precision tools, fixtures, gauges, and other measuring devices that are used to cut or form metal and other manufacturing materials. Die makers construct dies or molds for metal, plastics, ceramics and other products. Both work from blueprints to visualize and manufacture the products from start to finish.

Get Started Now!

- Find a tool- and die-making manufacturing company in your area. Ask to set up an informational interview to get a tour of the grounds and working environment.
- Talk to a tool and die maker about his or her experience in the field. You can find companies in your local phone book.
- Take science, math, and machine shop. This knowledge will allow you to start an apprenticeship program with a manufacturing company.

Search It!
National Tooling and Machining Association at ***www.ntma.org***

Read It!
Mold Maker magazine at ***www.moldmakermag.com***

Learn It!
Four- to five-year apprenticeship program, combining on-the-job training with classroom study is typical.

Earn It!
Median hourly wage is $20.54. (Source: U.S. Department of Labor)

Find It!
Tool and die makers work for machine shops and tool manufacturers. You can find a company in your area when you register (for free!) as a job seeker at ***www.tool-and-die-maker-jobs.com***.

Hire Yourself!

Learn your way around metalforming with a visit to the Precision Metalforming Association Educational Foundation website at ***www.pmaef.org/index.htm***. Use the information you find there to create a chart that illustrates the various metalforming careers and specializations. Make sure to distinguish between entry-level positions and those requiring additional training and experience.

First, they figure out what steps to take to manufacture a product. Next, tool and die makers measure and mark the metal pieces using protractors, rulers, and other devices. Later, they set up the tools and machines required for the job. Upon loading the materials—metals, plastics, or ceramics—they have to make sure the parts are shaped and ground according to the blueprints. Next, they smooth, polish, and shape the components using grinders, stones, files, and scrapers. After they verify the exact dimensions of the parts, they get to work assembling the product with bolts, glue, and hand tools. Depending on the product, they may place it in a specially-heated furnace to harden. Finally—when the product looks finished—they make any necessary nips, tucks, or repairs until it is perfect. One look at these steps, and you'll understand why tool and die makers have to be patient, precise, and focused on perfection.

If you're considering this field, you may want to pay attention in science class. In addition to operating machinery, tool and die makers have to know the chemical and physical properties of metals and other materials. In other words, they have to know how hard or soft a substance is and how it will react to heating, stretching, shaping, and pounding. Computer skills are also essential. Tool and die makers now use computer-aided design (CAD) to develop products and parts. They have to know the ins and outs of CAD programs and the machines that use them.

If this sound like the field for you, you're in luck. According to the Bureau of Labor Statistics, tool and die makers will enjoy excellent career opportunities—with little competition. The number of job openings will outnumber the number of workers trained in this field. And that means highly skilled workers will be able to handpick their perfect job. So, where do potential tool and die makers—like yourself—find a job? Almost anywhere. Most of the companies in the U.S. metalworking industry have traditionally been located in the Midwest, Northeast, and West, but tool- and die-making shops can be found throughout the country.

welder Do you want a career that will go down in history? Become a welder! Welders have played a major role in the manufacturing history of the United States. In fact, without the skills and innovation of welders and welding companies, our world would be a much different place today.

Welders join two or more pieces of metal together using heat. In the United States welding began its rise to fame during World War II. That's when shipbuilders replaced rivets with welding. Rivets were used to hold together two or more pieces of metal. Manufacturers and the federal government realized that welding would significantly cut down the time it took to build a ship. They also saw that welded metals stayed together better than metals held by rivets. That launched welding into a host of national manufacturing projects, such as the Alaskan pipeline and famous high-rise buildings—and the need for welders rose dramatically.

Today, welder jobs continue to grow—with the U.S. Department of Labor projecting an excellent job outlook through the year 2012. Modern welders are still involved in large federal projects, such as the space shuttle program and satellites. But more and more, welders are

Get Started Now!

- Take machine shop courses while in high school. You'll be exposed to various tools, including welding and soldering equipment.
- Find a local welder and inquire about apprenticeship opportunities. Welders who own their own shop may welcome student help and offer valuable career advice.
- If there's a power plant near you, plan a visit. Call in advance to set up an appointment with a welder who works there. Ask for a tour of the grounds and a description of a typical day on the job.

Search It!
American Welding Society at ***www.aws.org***

Read It!
Read about a minitool created for a big job in *BusinessWeek Online* at ***www.businessweek.com/magazine/content/02_40/c3802114.htm***.

Learn It!
On-the-job training plus community college or technical school training are typical.

Earn It!
Median hourly wage is $14.02. (Source: U.S. Department of Labor)

Find It!
Welders can find work in a variety of places, including mines, industrial sites, mechanic shops, shipyards, logging camps, or any place where welding is required. Con Edison of New York (***www.coned.com***) hires welders, as do other energy companies.

Hire Yourself!

What do the Golden Gate Bridge, the St. Louis Arch, and a NASA liftoff platform all have in common? They are great examples of the wonders of welding. Go on-line to find photographs of examples of products that were welded. Hints of things to look for include structural buildings and bridges, bank doors and vaults, jail cells, off-shore drilling platforms, cranes, and geodesic domes. Use the photographs to create a "wonders of welding" collage.

finding jobs in private industry, including construction, shipyards, and mechanic shops.

What does it take to be a welder? First, welders can't be apprehensive around heat and fire. For the majority of the day, welders handle and operate high-pressure torches and other flame-cutting equipment. Second, they must be able to endure less-than-perfect working conditions. Welders usually work in shops that are hot, smoky, and smell of burning metals. Third, welders must have good hand-eye coordination and be mechanically inclined—at all angles. The job requires welding pieces in flat, horizontal, vertical, or overhead positions. Dexterity is a must.

Welders also have to decide what type of welding they will do. Technical schools or vocational-technical high school programs can introduce prospective welders to the various welding techniques. Experts in the field say trade schools and community colleges offer the best opportunities to learn the trade. Additionally, long-distance training may be another avenue if technical schools are not in your area. On-the-job training or apprenticeship programs are also key: they provide a combination of real-life training beside an experienced welder coupled with classroom instruction. And learning doesn't end once you've become experienced. Welders must keep abreast of technology changes, especially in automation. Automated machines have changed the welding workplace, and welders will do their career a favor by learning how to use and maintain automated equipment.

If you're interested in welding, you may also want to look into a career as a cutter, solderer, or brazier.

woodworker

The wood industry offers a host of career opportunities, ranging from science-based careers in forestry, to manufacturing careers in carpentry, to industrial careers in sawmills—and everything in between. People who enter the wood industry can base their career path on their individual interests and the type of schooling they'd like to complete. Some choose to work in research or management for state or federal forestry services, construction companies, logging companies, or investment companies. Others become independent consultants, working on diverse projects as they become available.

Wood science programs offered through four-year colleges and universities prepare students for forestry-related bachelor's or master's degrees. The course workload is fairly intense, including biology, chemistry, physics, ecology, math, business management, and communications classes. Students who enroll in wood science programs take in-depth classes about wood. Some schools even require hands-on prepping in which students spend a semester or more in a forestry center before being admitted to the wood science program.

Get Started Now!

- Take wood shop courses in high school, no matter which career you're interested in. It will give you a solid foundation in learning about wood types and characteristics.
- Take math, science, and computer courses. All woodworking careers require some level of these skills.
- Join your local 4-H Club, Future Farmers of America Club, or student council. You'll pick up leadership skills, as well as learn about environmental issues.
- Check out the wood-related chat sessions—and other wood websites—at ***www.woodworking.org***.

Search It!
Woodworkers Website Association at ***www.woodworking.org***

Read It!
Woodworker's Gazette at ***www.woodworking.org/WC/gazette.html***

Learn It!
Attend a technical woodworking school or enroll in a college or university wood science program.

Earn It!
Median hourly wage is $11.54. (U.S. Department of Labor)

Find It!
Woodworkers can find work in a variety of places, including logging and construction companies, sawmills, universities, state and federal forestry services, furniture and woodworking companies, and small wood shops.

Hire Yourself!

What better way to work with wood than, well, working with wood? Head to your local library, bookstore, or ***www.am-wood.com***, and pick up a book of woodworking plans. Choose a project that suits your skill level, tools, time, and budget. As you work, keep a journal of your work experience: what obstacles you have encountered, what you have learned, helpful tips and techniques, mistakes to avoid, etc. Share your journal—and finished project—with friends or family.

If you'd rather be in a hands-on job—or if the idea of four-plus years of college doesn't interest you—a career in woodworking has lots to offer. If you're great at making stuff, this is an excellent opportunity to build on your skills. Some woodworkers own or work in small wood shops and create custom products for customers. Cabinet companies, furniture companies, and others hire workers to create and manufacture their products. You might not have control over what you make—cabinets and furniture are built to exact specifications—but you'll take pride in knowing you built a product from scratch.

If you like the idea of working with wood, but can't turn a block of it into a functional piece of furniture, don't despair. There's room for

you, too, in the woodworking industry. (Didn't we say woodworking has a lot to offer?) Some technical schools have programs that train students for positions in sawmills, lumber companies, and other manufacturing industries. Some schools offer co-op programs with wood industry companies that allow students to earn while they learn. Most likely, you'll start out doing the hauling work, but it's a great opportunity to learn the ropes.

No matter the specialization, all woodworkers are employed at some stage of the process through which logs of wood are transformed into finished products. Standard work gear is required for almost all woodworking careers: hardhat, steel-toed boots, rugged and durable clothing, and safety goggles. You are, after all, working with huge trees, sawdust, tools, and machines. Woodworkers who work outdoors also have to pile on the sunscreen and heat protection gear. Part of the work gear is a fully-loaded toolbox. Woodworkers use hammers, chisels, saws, miter boxes, and other woodcutting and sanding equipment to do their work. Be prepared to spend a significant amount of money on the toolbox.

Big Question #5: do you have the right skills?

Career exploration is, in one sense, career matchmaking. The goal is to match your basic traits, interests and strengths, work values, and work personality with viable career options.

But the "stuff" you bring to a job is only half of the story.

Choosing an ideal job and landing your dream job is a two-way street. Potential employers look for candidates with specific types of skills and backgrounds. This is especially true in our technology-infused, global economy.

In order to find the perfect fit, you need to be fully aware of not only what you've got, but also what prospective employers need.

The following activity is designed to help you accomplish just that. This time we'll use the "wannabe" approach —working with careers you think you want to consider. This same matchmaking process will come in handy when it comes time for the real thing too.

Unfortunately, this isn't one of those "please turn to the end of the chapter and you'll find all the answers" types of activities. This one requires the best critical thinking, problem-solving, and decision-making skills you can muster.

Big Activity #5:
do you have the right skills?

Here's how it works:

Step 1: First, make a chart like the one on page 130.

Step 2: Next, pick a career profile that interests you and use the following resources to compile a list of the traits and skills needed to be successful. Include:
- Information featured in the career profile in this book;
- Information you discover when you look through websites of any of the professional associations or other resources listed with each career profile;
- Information from the career profiles and skills lists found on-line at America's Career InfoNet at ***www.acinet.org***.

Briefly list the traits or skills you find on separate lines in the first column of your chart.

Step 3: Evaluate yourself as honestly as possible. If, after careful consideration, you conclude that you already possess one of the traits or skills included on your list, place an *X* in the column marked "Got It!" If you conclude that the skill or trait is one you've yet to acquire, follow these directions to complete the column marked "Get It?":
- If you believe that gaining proficiency in a skill is just a matter of time and experience and you're willing to do whatever it takes to acquire that skill, place a *Y* (for yes) in the corresponding space.
- Or, if you are quite certain that a particular skill is one that you don't possess now, and either can't or won't do what it takes to acquire it, mark the corresponding space with an *N* (for no). For example, you want to be a brain surgeon. It's important, prestigious work and the pay is good. But, truth be told, you'd rather have brain surgery yourself than sit through eight more years of really intense science and math. This rather significant factor may or may not affect your ultimate career choice. But it's better to think it through now rather than six years into med school.

Step 4: Place your completed chart in your Big Question AnswerBook.

When you work through this process carefully, you should get some eye-opening insights into the kinds of careers that are right for you. Half reality check and half wake-up call, this activity lets you see how you measure up against important workforce competencies.

Big Activity #5: **do you have the right skills?**

skill or trait required	**got it!**	**get it!**

more career ideas in manufacturing

Careers featured in the previous section represent mainstream, highly viable occupations where someone with the right set of skills and training stands more than half a chance of finding gainful employment. However, these ideas are just the beginning. There are lots of ways to make a living in any industry—and this one is no exception.

Following is a list of career ideas related in one way or another to the manufacturing industry. This list is included here for two reasons. First, to illustrate some unique ways to blend your interests with opportunities. Second, to keep you thinking beyond the obvious.

As you peruse the list you're sure to encounter some occupations you've never heard of before. We hope you get curious enough to look them up. Others may trigger one of those "aha" moments where everything clicks and you know you're onto something good. Either way we hope it helps point the way toward some rewarding opportunities in the manufacturing.

Automated Manufacturing Technician

Biomedical Equipment Technician

Bookbinder

Calibration Technician

Communication System Installer

Communication System Repairer

Communications Manager

Computer Installer

Computer Maintenance Technician

Computer Repairer

Design Engineer

Electrical Installer

Electrical Repairer

Electromechanical Equipment Assembler

Electronic Technician

Electronic Technologist

Engineering Technician

Environmental Specialist

Extruding and Drawing Machine Setter

Extrusion Machine Operator

Facility Electrician

Fixture Designer

Foundry Worker

Freight Mover

Grinding, Lapping, and Buffing Machine Operator

Health and Safety Manager

Industrial Electronic Technician

Industrial Facilities Manager

Industrial Machinery Mechanic

Industrial Maintenance Electrician

Industrial Truck and Tractor Operator

Instrument Calibration Technician

Instrument Control Technician

Instrument Worker

Lab Technician

Laser Systems Technician

Logistical Engineer

Logistician

Machine Operator

Maintenance Repairer

Major Appliance Repairperson

Material Handler

Medical Appliance Maker

Meter Installer

Milling Machine Operator

Packager

Painter

Pipe Fitter

Plumber

Precision Inspector

Precision Layout Worker

Process Control Technician

Process Improvement Technician

Production Associate

Production Manager

Quality Engineer

Safety Coordinator

Safety Engineer

Safety Technician

Security System Installer

Security System Repairer

Set-up Operator

Shipping and Receiving Clerk

Traffic Manager

Transportation Manager

Utility Manager

Big Question #6: **are you on the right path?**

You've covered a lot of ground so far. You've had a chance to discover more about your own potential and expectations. You've taken some time to explore the realities of a wide variety of career opportunities within this industry.

Now is a good time to sort through all the details and figure out what all this means to you. This process involves equal measures of input from your head and your heart. Be honest, think big, and, most of all, stay true to you.

You may be considering an occupation that requires years of advanced schooling which, from your point of view, seems an insurmountable hurdle. What do you do? Give up before you even get started? We hope not. We'd suggest that you try some creative thinking.

Big Activity #6:
are you on the right path?

Start by asking yourself if you to want pursue this particular career so badly that you're willing to do whatever it takes to make it. Then stretch your thinking a little to consider alternative routes, nontraditional career paths, and other equally meaningful occupations.

Following are some prompts to help you sort through your ideas. Simply jot down each prompt on a separate sheet of notebook paper and leave plenty of space for your responses.

Big Activity #6: **are you on the right path?**

ARE YOU ON THE RIGHT PATH?

One thing I know for sure about my future occupation is

I'd prefer to pursue a career that offers

I'd prefer to pursue a career that requires

A career option I'm now considering is

What appeals to me most about this career is

What concerns me most about this career is

Things that I still need to learn about this career include

Big Activity #6: **are you on the right path?**

Another career option I'm considering is

What appeals to me most about this career is

What concerns me most about this career is

Things that I still need to learn about this career include

Of these two career options I've named, the one that best fits most of my interests, skills, values, and work personality is because

At this point in the process, I am

❑ Pretty sure I'm on the right track

❑ Not quite sure yet but still interested in exploring some more

❑ Completely clueless about what I want to do

ARE YOU ON THE RIGHT PATH?

SECTION 3

experiment with success

Right about now you may find it encouraging to learn that the average person changes careers five to seven times in his or her life. Plus, most college students change majors several times. Even people who are totally set on what they want to do often end up being happier doing something just a little bit different from what they first imagined.

So, whether you think you've found the ultimate answer to career happiness or you're just as confused as ever, you're in good company. The best advice for navigating these important life choices is this: Always keep the door open to new ideas.

As smart and dedicated as you may be, you just can't predict the future. Some of the most successful professionals in any imaginable field could never ever have predicted what—and how—they would be doing what they actually do today. Why? Because when they were in high school those jobs didn't even exist. It was not too long ago that there were no such things as personal computers, Internet research, digital cameras, mass e-mails, cell phones, or any of the other newfangled tools that are so critical to so many jobs today.

Keeping the door open means being open to recognizing changes in yourself as you mature and being open to changes in the way the world works. It also involves a certain willingness to learn new things and tackle new challenges.

It's easy to see how being open to change can sometimes allow you to go further in your chosen career than you ever dreamed. For instance, in almost any profession you can imagine, technology has fueled unprecedented opportunities. Those people and companies who have embraced this "new way of working" have often surpassed their original expectations of success. Just ask Bill Gates. He's now one of the world's wealthiest men thanks to a company called Microsoft that he cofounded while still a student at Harvard University.

It's a little harder to see, but being open to change can also mean that you may have to let go of your first dream and find a more appropriate one. Maybe your dream is to become a professional athlete. At this point in your life you may think that there's nothing in the world that would possibly make you happier. Maybe you're right and maybe you have the talent and persistence (and the lucky breaks) to take you all the way.

But maybe you don't. Perhaps if you opened yourself to new ideas you'd discover that the best career involves blending your interest in sports with your talent in writing to become a sports journalist or sports information director. Maybe your love of a particular sport and your interest in working with children might best be served in a coaching career. Who knows what you might achieve when you open yourself to all the possibilities?

So, whether you've settled on a career direction or you are still not sure where you want to go, there are several "next steps" to consider. In this section, you'll find three more Big Questions to help keep your career planning moving forward. These Big Questions are:

- Big Question #7: **who knows what you need to know?**
- Big Question #8: **how can you find out what a career is really like?**
- Big Question #9: **how do you know when you've made the right choice?**

Big Question #7: who knows what you need to know?

When it comes to the nitty-gritty details about what a particular job is really like, who knows what you need to know? Someone with a job like the one you want, of course. They'll have the inside scoop—important information you may never find in books or websites. So make talking to as many people as you can part of your career planning process.

Learn from them how they turned their own challenges into opportunities, how they got started, and how they made it to where they are now. Ask the questions that aren't covered in "official" resources, such as what it is really like to do their job, how they manage to do a good job and have a great life, how they learned what they needed to learn to do their job well, and the best companies or situations to start in.

A good place to start with these career chats or "informational interviews" is with people you know—or more likely, people you know who know people with jobs you find interesting. People you already know include your parents (of course), relatives, neighbors, friends' parents, people who belong to your place of worship or club, and so on.

All it takes to get the process going is gathering up all your nerve and asking these people for help. You'll find that nine and a half times out of 10, the people you encounter will be delighted to help, either by providing information about their careers or by introducing you to people they know who can help.

hints and tips for a successful interview

• TIP #1

Think about your goals for the interview, and write them down.

Be clear about what you want to know after the interview that you didn't know before it.

Remember that the questions for all personal interviews are not the same. You would probably use different questions to write a biography of the person, to evaluate him or her for a job, to do a history of the industry, or to learn about careers that might interest you.

Writing down your objectives will help you stay focused.

• TIP #2

Pay attention to how you phrase your questions.

Some questions that we ask people are "closed" questions; we are looking for a clear answer, not an elaboration. "What time does the movie start?" is a good example of a closed question.

Sometimes, when we ask a closed question, we shortchange ourselves. Think about the difference between "What times are the showings tonight?" and "Is there a 9 P.M. showing?" By asking the second question, you may not find out if there is an 8:45 or 9:30 show.

That can be frustrating. It usually seems so obvious when we ask a question that we expect a full answer. It's important to remember, though, that the person hearing the question doesn't always have the same priorities or know why the question is being asked.

The best example of this? Think of the toddler who answers the phone. When the caller asks, "Is your mom home?" the toddler says, "Yes" and promptly hangs up. Did the child answer the question? As far as he's concerned, he did a great job!

Another problem with closed questions is that they sometimes require so many follow-up questions that the person being interviewed feels like a suspect in an interrogation room.

A series of closed questions may go this way:

Q: What is your job title?
A: Assistant Producer
Q: How long have you had that title?
A: About two years.

Q: What was your title before that?
Q: How long did you have that title?
Q: What is the difference between the two jobs?
Q: What did you do before that?
Q: Where did you learn to do this job?
Q: How did you advance from one job to the next?

An alternative, "open" question invites conversation. An open-question interview might begin this way:

I understand you are an Assistant Producer. I'm really interested in what that job is all about and how you got to be at the level you are today.

Open questions often begin with words like:

Tell me about . . .
How do you feel about . . .
What was it like . . .

• TIP #3

Make the person feel comfortable answering truthfully.

In general, people don't want to say things that they think will make them look bad. How to get at the truth? Be empathic, and make their answers seem "normal."

Ask a performer or artist how he or she feels about getting a bad review from the critics, and you are unlikely to hear, "It really hurts. Sometimes I just want to cry and get out of the business." Or "Critics are so stupid. They never understand what I am trying to do."

Try this approach instead: "So many people in your industry find it hard to deal with the hurt of a bad critical review. How do you handle it when that happens?"

ask the experts

You can learn a lot by interviewing people who are already successful in the types of careers you're interested in. In fact, we followed our own advice and interviewed several people who have been successful in the field of manufacturing to share with you here.

Before you get started on your own interview, take a few minutes to look through the results of some of ours. To make it easier for you to compare the responses of all the people we interviewed, we have presented our interviews as a panel discussion that reveals important "success" lessons these people have learned along the way. Each panelist is introduced on the next page.

Our interviewees gave us great information about things like what their jobs are really like, how they got to where they are, and even provided a bit of sage advice for people like you who are just getting started.

So Glad You Asked

In addition to the questions we asked in the interviews in this book, you might want to add some of these questions to your own interviews:

- How did your childhood interests relate to your choice of career path?
- How did you first learn about the job you have today?
- In what ways is your job different from how you expected it to be?
- Tell me about the parts of your job that you really like.
- If you could get someone to take over part of your job for you, what aspect would you most like to give up?
- If anything were possible, how would you change your job description?
- What kinds of people do you usually meet in your work?
- Walk me through the whole process of getting your type of product made and distributed. Tell me about all the people who are involved.
- Tell me about the changes you have seen in your industry over the years. What do you see as the future of the industry?
- Are there things you would do differently in your career if you could do it all over?

real people with real jobs in manufacturing

Following are introductions to our panel of experts. Get acquainted with their backgrounds and then use their job titles to track their stories throughout the five success lessons.

- **Mark Fitzgerald** is a **Senior Industry Analyst** and is based in Boston, Massachusetts. His employer is Strategy Analytics, a consulting company specializing in strategic and tactical support for business planners around the world.
- **Diana Gubitosi** is a **Purchasing Agent** at a textile manufacturing company. She lives in Howell, New Jersey.
- **Aly Khalifa** is a **Product Designer** for Gamil Design in Raleigh, North Carolina. He specializes in designing and engineering sporting goods equipment.
- **Nichol Mackey** is **Operations Manager** for Airpacks, Inc., a company that designs and manufactures ergonomic backpacks. She's also involved in production development for the company.
- **John Potter, Jr., Tool and Die Manufacturer,** is the president of CAP Collet and Tool, a tool and die manufacturing company in Wyandotte, Michigan.
- **Ray Rocker** is **Production Manager** for Jelly Belly Candy Company, manufacturers of over 50 official "taste defining" jelly beans. He works in Fairfield, California.
- **Dale Senatore** is an **NDT** (nondestructive testing) **Engineer**. He lives in Lexington, Kentucky, where he manages an NDT and materials laboratory for the federal government.

Mark Fitzgerald

Aly Khalifa

Nichol Mackey

Ray Rocker

Success Lesson #1:
Work is a good thing when you find the right career.

- **Tell us what it's like to work in your current career.**

NDT Engineer: At our facility we basically perform the same type test on materials that you may undergo at a hospital. We x-ray the material to verify it does not contain discontinuities such as gas pockets or cracks. We conduct a chemical analysis test to verify the correct alloying elements are present to make up the alloy specified. We conduct hardness tests and numerous other forms of testing as well. We also verify the condition of aircraft parts and components removed from aircraft at scheduled intervals in order to reuse the parts. Whenever we recertify aircraft parts we are required to perform at least two methods of nondestructive testing. The methods used can be any combination of the following: X ray or ultrasound, followed by dye penetration, magnetic particle, or eddy current. Visual inspection is always included as a method and may include simply looking at the parts or complicated bore scope inspections.

Operations Manager: Being in product development, I create products for our backpack line. In order to do this, I research current trends as they relate to color and style and I incorporate them into our product line. I work closely with designers who draw our visions. From there, the drawings are sent to our partner in China where samples are created.

Production Manager: I manage the production of Jelly Belly candy, a very popular candy that is sold all over the world. In the 36 years I've been in this business I've made hundreds of different flavors and colors of jelly beans. Sometimes I've blended some of them together to come up with a different taste.

Purchasing Agent: My job as purchasing agent involves communicating with all our textile mills on a technical level about what I would like to achieve in the final products.

Tool and Die Manufacturer: I am the owner of a small business that supplies special tools to General Motors, Ford, and Chrysler.

Everybody has to start some-where!

Following is a list of first jobs once held by our esteemed panel of experts.

Soldier

Landscaper

Cook

Waiter

Freelance logo designer

Construction worker

Police officer

Snow shoveler

Sales clerk

Farm worker

Success Lesson #2:
Career goals change and so do you

- **What career did you hope to pursue when you were in high school?**

NDT Engineer: When I was in high school I wanted to be a football coach and history teacher. I had no idea of what a nondestructive testing (NDT) engineer was.

Operations Manager: I never had a focus when I was a young adult.

Product Designer: I'm pretty much doing what I've always wanted to do. The only difference is that I thought I'd have more time to design. I didn't know that I would have to spend so much time communicating with others.

Production Manager: As a teenager growing up in the South I just wanted to get through school and get a good job where I could have a little influence over my work.

Purchasing Agent: I wanted to be a schoolteacher or nurse.

Senior Industry Analyst: I didn't know at the time so I tried to expose myself to many options. However, I have always been interested in cars and all mechanical things.

Tool and Die Manufacturer: As a teenager, I always wanted to be in law enforcement.

- **What was it that made you change directions?**

NDT Engineer: I attended Arizona Western College for about a year to obtain my goal but partied too hard and could not keep my grades up. Uncle Sam had the cure back in those days—I got drafted into the army. I learned to like the army. I got assigned to duties as an aircraft mechanic and got sent to Vietnam for a couple of tours, but life was good. My formal training was in aerospace where I attended Embry-Riddle Aeronautical University for aircraft maintenance while in the army. While in the army I was also sent to several schools where I was certified in the basic methods of NDT. I never felt for a moment that I would later be earning a very handsome living using this training.

Operations Manager: I started off as an administrative assistant at a computer company. In the business world it seems I found what I was good at and stuck with it. Operations and customer service are rewarding in that I solve many problems and I feel I have accomplished something at the end of the day.

Production Manager: I was working for my uncle's construction company and we were repairing the floor of a little candy factory. They were looking for people who were willing to work and grow with the company. So I took advantage of that opportunity to start a career in candy-making. That was over 36 years ago and I'm really glad that I made that decision all those years ago.

Tool and Die Manufacturer: I was a policeman for seven years. When I was laid off I joined my father and his associates at CAP Collet.

Success Lesson #3: **One thing leads to another along any career path.**

- **How did you end up doing what you're doing now?**

NDT Engineer: While on a trip to Texas two things took place that changed my future. First, I was sent to a sheet metal manufacturer to get parts for our next aircraft. The ones we ordered were not ready. However, they had some parts in stock that were identical to our drawings so I signed to purchase them even though I was not authorized to do so.

Then I went to visit our parent company, which was also located in Texas. My purpose there was to take photographs of the NDT facility for the purpose of training the inspectors back in Lexington. While there I learned that they were trying to sell some older NDT equipment and I talked them into giving it to our contract—another decision I was not authorized to make.

The next day I went to a Chinese restaurant and my fortune cookie said that I would be going into a new career field. This prediction proved to be true because when I got back to Lexington my manager fired me for signing those unauthorized contracts.

I worked on other projects for a year or two but eventually convinced my managers that having a good NDT and materials laboratory would be a good thing for our company. They finally gave me the green light and after a couple of years we put together a state-of-the-art facility.

Operations Manager: I started out as the operations manager for Airpacks three years ago. I am a great problem solver so operations has always been what I do best. Last summer the owner of Airpacks asked if I would be interested in helping her with product development. I reminded her that I have no formal training but she felt that I was more than competent to do the job. So she started training me. Several months later I am now working on our 2005 backpack line. This has been a wonderful learning experience for me and it's also great for my résumé.

Product Designer: Starting my own company was a big risk, but it did pay off. I would have had to move every two years otherwise. My first full time job introduced me to many new things. By keeping up with those relationships, I have been able to grow with them even though I left that job years ago.

Purchasing Agent: I needed a job after working as medical assistant for an orthopedic doctor. I found a job as a file clerk in the garment district through an employment agency about thirty years ago. I always loved fashion and the company manufactured textiles. They taught me all I know through experience and on-the-job training.

Senior Industry Analyst: One of my first adult jobs was in automobile insurance. Even though I already knew about the mechanics of a car I didn't know all of the businesses that support the auto industry. This job provided a glimpse into other aspects of the automobile industry that I found really interesting.

Success Lesson #4: There's more than one way to get an education.

- **Where did you learn the skills of your field, both formally (school) and informally (experience)?**

NDT Engineer: I joined the American Society for Materials and made several friends within the local Bluegrass Chapter who were either degreed materials engineers or professors from the

University of Kentucky's Materials Department. I knew and understood the nondestructive testing aspects of my work but lacked the knowledge of the actual materials themselves. I used the Bluegrass Chapter as a networking system to fill in the voids. I owe a lot to these members for all the technical information they provided me in the early years of setting up the materials portion of the laboratory.

Product Designer: If you are involved in manufacturing, you must go to the factories. They are the heart and soul of the industrial economy. It really makes you confront the reality of the modern world.

Production Manager: I had a very good trainer when it came to making candy. He was the best teacher in the business and I worked very hard to learn this trade. I also went to the best confectionery school where I learned a lot about making candy. After that it was practice and trial and error, which has resulted in lots of success. I think the hands-on experiences are the best teacher you could have for success at any job.

Senior Industry Analyst: I have always been a curious person. If I don't know what a word means, I look it up. You don't have to know everything. It's more important to know how to get answers.

Success Lesson #5:
Good choices and hard work are a potent combination.

- **What are you most proud of in your career?**

NDT Engineer: I am currently on the Faculty Advisory Committee for the University of Kentucky Material Science Department for Industry. I've also been nominated to the International Committee for Welding Aerospace Applications and recently attended my first meeting at the Kennedy Space Center in Cape Canaveral, Florida.

Operations Manager: The fact that I have worked my way up the corporate ladder with no college degree is fulfilling. Not to say that one doesn't need a degree; I believe school is very important. I have always been a "hands-on" type of person where learning is better if I am thrown into a situation as opposed to sitting in front of a textbook.

Product Designer: We formed a consortium called Designbox (www.designbox.us) where independent designers can share ideas and resources. It's been eye-opening.

Production Manager: I enjoy having the skills to do all the jobs in my line of work, making candy from the beginning all the way through the process. I made the very first Jelly Belly Beans back in the mid 1970s. It was lots of fun because Jelly Bellys happened to be President Ronald Reagan's favorite candy. He kept a jar on his desk and always liked to brag about them.

Purchasing Agent: My ability to listen and learn on the job.

Tool and Die Manufacturer: Being part of a business that three men (including author Diane Lindsey Reeves' grandfather, Raymond Austin) took a chance on over 40 years ago is especially satisfying. They did the hard work of building it; now all I have to do is maintain it.

Big Activity #7:
who knows what you need to know?

It's one thing to read about conducting an informational interview, but it's another thing altogether to actually do one. Now it's your turn to shine. Just follow these steps for doing it like a pro!

Step 1: Identify the people you want to talk to about their work.
Step 2: Set up a convenient time to meet in person or talk over the phone.
Step 3: Make up a list of questions that reflect things you'd really like to know about that person's work. Go for the open questions you just read about.
Step 4: Talk away! Take notes as your interviewee responds to each question.
Step 5: Use your notes to write up a "news" article that describes the person and his or her work.
Step 6: Place all your notes and the finished "news" article in your Big Question AnswerBook.

Big Activity #7: **who knows what you need to know?**

contact information	appointments/sample questions
name company title address phone email	day time location sample questions:
name company title address phone email	day time location sample questions:
name company title address phone email	day time location sample questions:

CONTACT INFO

Big Activity #7: **who knows what you need to know?**

questions	answers
	INTERVIEW NOTES

Big Activity #7: **who knows what you need to know?**

questions	answers
	INTERVIEW NOTES

Big Activity #7: **who knows what you need to know?**

NEWS

Big Activity #7: **who knows what you need to know?**

NEWS

?

Big Question #8: how can you find out what a career is really like?

There are some things you just have to figure out for yourself. Things like whether your interest in pursuing a career in marine biology is practical if you plan to live near the Mojave Desert.

Other things you have to see for yourself. Words are sometimes not enough when it comes to conveying what a job is really like on a day-to-day basis—what it looks like, sounds like, and feels like.

Here are a few ideas for conducting an on-the-job reality check.

identify typical types of workplaces

Think of all the places that jobs like the ones you like take place. Almost all of the careers in this book, or ones very similar to them, exist in the corporate world, in the public sector, and in the military. And don't forget the option of going into business for yourself!

For example: Are you interested in public relations? You can find a place for yourself in almost any sector of our economy. Of course, companies definitely want to promote their products. But don't limit yourself to the Fortune 500 corporate world. Hospitals, schools, and manufacturers need your services. Cities, states, and even countries also need your services. They want to increase tourism, get businesses to relocate there, and convince workers to live there or students to study there. Each military branch needs to recruit new members and to show how they are using the money they receive from the government for medical research, taking care of families, and other non-news-breaking uses. Charities, community organizations, and even religious groups want to promote the good things they are doing so that they will get more members, volunteers, contributions, and funding. Political candidates, parties, and special interest groups all want to promote their messages. Even actors, dancers, and writers need to promote themselves.

Not interested in public relations but know you want a career that involves lots of writing? You've thought about becoming the more obvious choices—novelist, newspaper reporter, or English teacher. But you don't want to overlook other interesting possibilities, do you?

What if you also enjoy technical challenges? Someone has to write the documentation for all those computer games and software.

Love cars? Someone has to write those owner's manuals too.

Ditto on those government reports about safety and environmental standards for industries.

Maybe community service is your thing. You can mix your love for helping people with writing grant proposals seeking funds for programs at hospitals, day care centers, or rehab centers.

Talented in art and design? Those graphics you see in magazine advertisements, on your shampoo bottle, and on a box of cereal all have to be created by someone.

That someone could be you.

find out about the job outlook

Organizations like the U.S. Bureau of Labor Statistics spend a lot of time and energy gathering data on what kinds of jobs are most in demand now and what kinds are projected to be in demand in the future. Find out what the job outlook is for a career you like. A good resource for this data can be found on-line at America's Career InfoNet at ***www.acinet.org/acinet.***

This information will help you understand whether the career options you find most appealing are viable. In other words, job outlook data will give you a better sense of your chances of actually finding gainful employment in your chosen profession—a rather important consideration from any standpoint.

Be realistic. You may really, really want to be a film critic at a major newspaper. Maybe your ambition is to become the next Roger Ebert.

Think about this. How many major newspapers are there? Is it reasonable to pin all your career hopes on a job for which there are only about 10 positions in the whole country? That doesn't mean that it's impossible to achieve your ambition. After all, someone has to fill those positions. It should just temper your plans with realism and perhaps encourage you to have a back-up plan, just in case.

look at training requirements

Understand what it takes to prepare yourself for a specific job. Some jobs require only a high school diploma. Others require a couple of years of technical training, while still others require four years or more in college.

Be sure to investigate a variety of training options. Look at training programs and colleges you may like to attend. Check out their websites to see what courses are required for the major you want. Make sure you're willing to "do the time" in school to prepare yourself for a particular occupation.

see for yourself

There's nothing quite like seeing for yourself what a job is like. Talk with a teacher or guidance counselor to arrange a job-shadowing opportunity with someone who is in the job or in a similar one.

Job shadowing is an activity that involves actually spending time at work with someone to see what a particular job is like up close and personal. It's an increasingly popular option and your school may participate in specially designated job-shadowing days. For some especially informative resources on job shadowing, visit ***www.jobshadow.org***.

Another way to test-drive different careers is to find summer jobs and internships that are similar to the career you hope to pursue.

make a Plan B

Think of the alternatives! Often it's not possible to have a full-time job in the field you love. Some jobs just don't pay enough to meet the needs of every person or family. Maybe you recognize that you don't have the talent, drive, or commitment to rise to the top. Or, perhaps you can't afford the years of work it takes to get established or you place a higher priority on spending time with family than that career might allow.

If you can see yourself in any of those categories, DO NOT GIVE UP on what you love! There is always more than one way to live out your dreams. Look at some of the other possibilities in this book. Find a way to integrate your passion into other jobs or your free time.

Lots of people manage to accomplish this in some fairly impressive ways. For instance, the Knicks City Dancers, known for their incredible performances and for pumping up the crowd at Knicks basketball games, include an environmental engineer, a TV news assignment editor, and a premed student, in addition to professional dancers. The Broadband Pickers, a North Texas bluegrass band, is made up of five lawyers and one businessman. In fact, even people who are extremely successful in a field that they love find ways to indulge their other passions. Paul Newman, the actor and director, not only drives race cars as a hobby, but also produces a line of gourmet foods and donates the profits to charity.

Get the picture? Good. Hang in there and keep moving forward in your quest to find your way toward a great future.

Big Activity #8:

how can you find out what a career is really like?

This activity will help you conduct a reality check about your future career in two ways. First, it prompts you to find out more about the nitty-gritty details you really need to know to make a well-informed career choice. Second, it helps you identify strategies for getting a firsthand look at what it's like to work in a given profession—day in and day out.

Here's how to get started:

Step 1: Write the name of the career you're considering at the top of a sheet of paper (or use the following worksheets if this is your book).

Step 2: Create a checklist (or, if this is your book, use the one provided on the following pages) covering two types of reality-check items.

First, list four types of information to investigate:

- training requirements
- typical workplaces
- job outlook
- similar occupations

Second, list three types of opportunities to pursue:

- job shadowing
- apprenticeship
- internship

Step 3: Use resources such as America's Career InfoNet at ***www.acinet.org*** and Career OneStop at ***www.careeronestop.org*** to seek out the information you need.

Step 4: Make an appointment with your school guidance counselor to discuss how to pursue hands-on opportunities to learn more about this occupation. Use the space provided on the following worksheets to jot down preliminary contact information and a brief summary of why or why not each career is right for you.

Step 5: When you're finished, place these notes in your Big Question AnswerBook.

Big Activity #8: **how can you find out what a career is really like?**

career choice:	
training requirements	
typical workplaces	
job outlook	
similar occupations	

INFORMATION

Big Activity #8: **how can you find out what a career is really like?**

job shadowing	when: where: who: observations and impressions:
apprenticeship	when: where: who: observations and impressions:
internship	when: where: who: observations and impressions:

OPPORTUNITIES

Big Question #9: how do you know when you've made the right choice?

When it comes right down to it, finding the career that's right for you is like shopping in a mall with 12,000 different stores. Finding the right fit may require trying on lots of different options.

All the Big Questions you've answered so far have been designed to expand your career horizons and help you clarify what you really want in a career. The next step is to see how well you've managed to integrate your interests, capabilities, goals, and ambitions with the realities of specific opportunities.

There are two things for you to keep in mind as you do this.

First, recognize the value of all the hard work you did to get to this point. If you've already completed the first eight activities thoughtfully and honestly, whatever choices you make will be based on solid knowledge about yourself and your options. You've learned to use a process that works just as well now, when you're trying to get an inkling of what you want to do with your life, as it will later when you have solid job offers on the table and need to make decisions that will affect your life and family.

Second, always remember that sometimes, even when you do everything right, things don't turn out the way you'd planned. That's called life. It happens. And it's not the end of the world. Even if you make what seems to be a bad choice, know this—there's no such thing as a wasted experience. The paths you take, the training you receive, the people you meet—they ultimately fall together like puzzle pieces to make you who you are and prepare you for what you're meant to do.

That said, here's a strategy to help you confirm that you are making the very best choices you can.

Big Activity #9:

how do you know when you've made the right choice?

One way to confirm that the choices you are making are right for you is to look at both sides of this proverbial coin: what you are looking for and what each career offers. The following activity will help you think this through.

Step 1: To get started, make two charts with four columns (or, if this is your book, use the following worksheet).

Step 2: Label the first column of the first chart as "Yes Please!" Under this heading list all the qualities you absolutely must have in a future job. This might include factors such as the kind of training you'd prefer to pursue (college, apprenticeship, etc.); the type of place where you'd like to work (big office, high-tech lab, in the great outdoors, etc.); and the sorts of people you want to work with (children, adults, people with certain needs, etc.). It may also include salary requirements or dress code preferences.

Step 3: Now at the top of the next three columns write the names of three careers you are considering. (This is a little like Big Activity #3 where you examined your work values. But now you know a lot more and you're ready to zero in on specific careers.)

Step 4: Go down the list and use an *X* to indicate careers that do indeed feature the desired preferences. Use an *O* to indicate those that do not.

Step 5: Tally up the number of *Xs* and *Os* at the bottom of each career column to find out which comes closest to your ideal job.

Step 6: In the first column of the second chart add a heading called "No Thanks!" This is where you'll record the factors you simply prefer not to deal with. Maybe long hours, physically demanding work, or jobs that require years of advanced training just don't cut it for you. Remember that part of figuring out what you do want to do involves understanding what you don't want to do.

Step 7: Repeat steps 2 through 5 for these avoid-at-all-costs preferences as you did for the must-have preferences above.

Big Activity #9: **how do you know when you've made the right choice?**

yes please!	career #1	career #2	career #3
totals	___X___O	___X___O	___X___O

Big Activity #9: **how do you know when you've made the right choice?**

no thanks!	career #1	career #2	career #3
totals	___X___O	___X___O	___X___O

Big Question #10: what's next?

Think of this experience as time well invested in your future. And expect it to pay off in a big way down the road. By now, you have worked (and perhaps wrestled) your way through nine important questions:

- Big Question #1: **who are you?**
- Big Question #2: **what are your interests and strengths?**
- Big Question #3: **what are your work values?**
- Big Question #4: **what is your work personality?**
- Big Question #5: **do you have the right skills?**
- Big Question #6: **are you on the right path?**
- Big Question #7: **who knows what you need to know?**
- Big Question #8: **how can you find out what a career is really like?**
- Big Question #9: **how do you know when you've made the right choice?**

But what if you still don't have a clue about what you want to do with your life?

Don't worry. You're talking about one of the biggest life decisions you'll ever make. These things take time.

It's okay if you don't have all the definitive answers yet. At least you do know how to go about finding them. The process you've used to work through this book is one that you can rely on throughout your life to help you sort through the options and make sound career decisions.

So what's next?

More discoveries, more exploration, and more experimenting with success are what come next. Keep at it and you're sure to find your way to wherever your dreams and ambitions lead you.

And, just for good measure, here's one more Big Activity to help point you in the right direction.

Big Activity #10: **what's next?**

List five things you can do to move forward in your career planning process (use a separate sheet if you need to). Your list may include tasks such as talking to your guidance counselor about resources your school makes available, checking out colleges or other types of training programs that can prepare you for your life's work, or finding out about job-shadowing or internship opportunities in your community. Remember to include any appropriate suggestions from the Get Started Now! list included with each career profile in Section 2 of this book.

Big Activity #10: **what's next?**

a final word

You are now officially equipped with the tools you need to track down a personally appropriate profession any time you have the need or desire. You've discovered more about who you are and what you want. You've explored a variety of career options within a very important industry. You've even taken it upon yourself to experiment with what it might be like to actually work in certain occupations.

Now it's up to you to put all this newfound knowledge to work for you. While you're at it, here's one more thing to keep in mind: Always remember that there's no such thing as a wasted experience. Certainly some experiences are more positive than others, but they all teach us something.

Chances are you may not get everything right the first time out. It may turn out that you were incorrect about how much you would love to go to a certain college or pursue a particular profession. That doesn't mean you're doomed to failure. It simply means that you've lived and learned. Sometimes you just won't know for sure about one direction or another until you try things out a bit. Nothing about your future has to be written in stone. Allow yourself some freedom to experiment with various options until you find something that really clicks for you.

Figuring out what you want to do with the rest of your life is a big deal. It's probably one of the most exciting and among the most intimidating decisions you'll ever make. It's a decision that warrants clear-headed thought and wholehearted investigation. It's a process that's likely to take you places you never dared imagine if you open yourself up to all the possibilities. Take a chance on yourself and seek out and follow your most valued hopes and dreams into the workplace.

Best wishes for a bright future!

Appendix

a virtual support team

As you continue your quest to determine just what it is you want to do with your life, you'll find that you are not alone. There are many people and organizations who want to help you succeed. Here are two words of advice—let them! Take advantage of all the wonderful resources so readily available to you.

The first place to start is your school's guidance center. There you are quite likely to find a variety of free resources which include information about careers, colleges, and other types of training opportunities; details about interesting events, job shadowing activities, and internship options; and access to useful career assessment tools.

In addition, since you are the very first generation on the face of the earth to have access to a world of information just the click of a mouse away—use it! The following Internet resources provide all kinds of information and ideas that can help you find your future.

make an informed choice

Following are five of the very best career-oriented websites currently on-line. Be sure to bookmark these websites and visit them often as you consider various career options.

America's Career Info Net ***www.acinet.org/acinet/default.asp***

Quite possibly the most comprehensive source of career exploration anywhere, this U.S. Department of Labor website includes all kinds of current information about wages, market conditions, employers, and employment trends. Make sure to visit the site's career video library where you'll find links to over 450 videos featuring real people doing real jobs.

Careers & Colleges ***www.careersandcolleges.com***

Each year Careers & Colleges publishes four editions of *Careers & Colleges* magazine, designed to help high school students set and meet their academic, career, and financial goals. Ask your guidance counselor about receiving free copies. You'll also want to visit the excellent Careers and Colleges website. Here you'll encounter their "Virtual Guidance Counselor," an interactive career database that allows you to match your interests with college majors or careers that are right for you.

Career Voyages ***www.careervoyages.gov***

This website is brought to you compliments of collaboration between the U.S. Department of Labor and the U.S. Department of Education and is designed especially for students like you. Here you'll find information on high-growth, high-demand occupations and the skills and education needed to attain those jobs.

Job Shadow *www.jobshadow.org*
See your future via a variety of on-line virtual job-shadowing videos and interviews featuring people with fascinating jobs.

My Cool Career *www.mycoolcareer.com*
This website touts itself as the "coolest career dream site for teens and 20's." See for yourself as you work your way through a variety of useful self-assessment quizzes, listen to an assortment of on-line career shows, and explore all kinds of career resources.

investigate local opportunities

To get a better understanding of employment happenings in your state, visit these state-specific career information websites.

Alabama
www.ajb.org/al
www.al.plusjobs.com

Alaska
www.jobs.state.ak.us
www.akcis.org/default.htm

Arizona
www.ajb.org/az
www.ade.state.az.us/cte/AZCrnproject10.asp

Arkansas
www.ajb.org/ar
www.careerwatch.org
www.ioscar.org/ar

California
www.calmis.ca.gov
www.ajb.org/ca
www.eurekanet.org

Colorado
www.coloradocareer.net
www.coworkforce.com/lmi

Connecticut
www1.ctdol.state.ct.us/jcc
www.ctdol.state.ct.us/lmi

Delaware
www.ajb.org/de
www.delewareworks.com

District of Columbia
www.ajb.org/dc
www.dcnetworks.org

Florida
www.Florida.access.bridges.com
www.employflorida.net

Georgia
www.gcic.peachnet.edu
(Ask your school guidance counselor for your school's free password and access code)
www.dol.state.ga.us/js

Hawaii
www.ajb.org/hi
www.careerkokua.org

Idaho
www.ajb.org/id
www.cis.idaho.gov

Illinois
www.ajb.org/il
www.ilworkinfo.com

Indiana
www.ajb.org/in
http://icpac.indiana.edu

Iowa
www.ajb.org/ia
www.state.ia.us/iccor

Kansas
www.ajb.org/ks
www.kansasjoblink.com/ada

Kentucky
www.ajb.org/ky

Louisiana
www.ajb.org/la
www.ldol.state.la.us/jobpage.asp

Maine
www.ajb.org/me
www.maine.gov/labor/lmis

Maryland
www.ajb.org/md
www.careernet.state.md.us

Massachusetts
www.ajb.org/ma
http://masscis.intocareers.org

Michigan
www.mois.org

Minnesota
www.ajb.org/mn
www.iseek.org

Mississippi
www.ajb.org/ms
www.mscareernet.org

Missouri
www.ajb.org/mo
www.greathires.org

Montana
www.ajb.org/mt
http://jsd.dli.state.mt.us/mjshome.asp

Nebraska
www.ajb.org/ne
www.careerlink.org

New Hampshire
www.nhes.state.nh.us

New Jersey
www.ajb.org/nj
www.wnjpin.net/coei

New Mexico
www.ajb.org/nm
www.dol.state.nm.us/soicc/upto21.html

Nevada
www.ajb.org/nv
http://nvcis.intocareers.org

New York
www.ajb.org/ny
www.nycareerzone.org

North Carolina
www.ajb.org/nc
www.ncsoicc.org
www.nccareers.org

North Dakota
www.ajb.org/nd
www.imaginend.com
www.ndcrn.com/students

Ohio
www.ajb.org/oh
https://scoti.ohio.gov/scoti_lexs

Oklahoma
www.ajb.org/ok
www.okcareertech.org/guidance
http://okcrn.org

Oregon
www.hsd.k12.or.us/crls

Pennsylvania
www.ajb.org/pa
www.pacareerlink.state.pa.us

Rhode Island
www.ajb.org/ri
www.dlt.ri.gov/lmi/jobseeker.htm

South Carolina
www.ajb.org/sc
www.scois.org/students.htm

South Dakota
www.ajb.org/sd

Tennessee
www.ajb.org/tn
www.tcids.utk.edu

Texas
www.ajb.org/tx
www.ioscar.org/tx
www.cdr.state.tx.us/Hotline/Hotline.html

Utah
www.ajb.org/ut
http://jobs.utah.gov/wi/occi.asp

Vermont
www.ajb.org/vt
www.vermontjoblink.com
www.vtlmi.info/oic.cfm

Virginia
www.ajb.org/va
www.vacrn.net

Washington
www.ajb.org/wa
www.workforceexplorer.com
www.wa.gov/esd/lmea/soicc/sohome.htm

West Virginia
www.ajb.org/wv
www.state.wv.us/bep/lmi

Wisconsin
www.ajb.org/wi
www.careers4wi.wisc.edu
http://wiscareers.wisc.edu/splash.asp

Wyoming
www.ajb.org/wy
http://uwadmnweb.uwyo.edu/SEO/wcis.htm

get a job

Whether you're curious about the kinds of jobs currently in big demand or you're actually looking for a job, the following websites are a great place to do some virtual job-hunting.

America's Job Bank ***www.ajb.org***

Another example of your (or, more accurately, your parent's) tax dollars at work, this well-organized website is sponsored by the U.S. Department of Labor. Job seekers can post resumes and use the site's search engines to search through over a million job listings by location or by job type.

Monster.com ***www.monster.com***

One of the Internet's most widely used employment websites, this is where you can search for specific types of jobs in specific parts of the country, network with millions of people, and find useful career advice.

explore by special interests

An especially effective way to explore career options is to look at careers associated with a personal interest or fascination with a certain type of industry. The following websites help you narrow down your options in a focused way.

What Interests You? ***www.bls.gov/k12***

This Bureau of Labor Statistics website provides information about careers associated with 12 special interest areas: math, reading, science, social studies, music and arts, building and fixing things, helping people, computers, law, managing money, sports, and nature.

Construct My Future ***www.constructmyfuture.com***

With over $600 billion annually devoted to new construction projects, about 6 million Americans build careers in this industry. This website, sponsored by the Association of Equipment Distributors, the Association of Equipment Manufacturers, and Associated General Contractors, introduces an interesting array of construction-related professions.

Dream It Do It ***www.dreamit-doit.com***

In order to make manufacturing a preferred career choice by 2010, the National Association of Manufacturing's Center for Workforce Success is reaching out to young adults and their parents, educators, communities, and policy-makers to change their minds about manufacturing's future and its careers. This website introduces high-demand 21st-century manufacturing professions many will find surprising and worthy of serious consideration.

Get Tech ***www.gettech.org***
Another award-winning website from the National Association of Manufacturing.

Take Another Look ***www.Nrf.com/content/foundation/rcp/main.htm***
The National Retail Federation challenges students to take another look at their industry by introducing a wide variety of careers associated with marketing and advertising, store management, sales, distribution and logistics, e-commerce, and more.

Index

Page numbers in **boldface** indicate main articles. Page numbers in *italics* indicate photographs.

C

D

E

F

G

H

I

T

U